CONTENTS

5	Pendants
15	Open-Type Bells
31	Military Items From The Napoleonic Period
43	Buckle Plates
65	Belt Chapes
103	Medieval Gilded Studs
105	Non-Heraldic Bells
109	Lead Weights

Roman bronze pendant
page 13

Tintinnabula suspension holder
page 28

Medieval lead weight **page 125**

Roman chape
page 70

Buckle and plate 1250-1500
page 62

Published by Greenlight Publishing,
The Publishing House, 119 Newland Street,
Witham, Essex CM8 1WF

Tel 01376 521900 **Fax** 01376 521901
books@greenlightpublishing.co.uk
www.greenlightpublishing.co.uk

Editor Greg Payne

Design Editor & Origination
Christine Jennett

All rights reserved. No part of this publication may be reproduced, stored in a retrieval system, or transmitted in any form by any means, electronic, mechanical photocopying, recording or otherwise, without the prior permission of Greenlight Publishing.

ISBN 978 1 897738 399

Printed in Great Britain
© Gordon Bailey 2011

Acknowledgements

Thanks go to my cousin David Bailey who is always willing to add his knowledge to mine on the subjects covered within these pages; I must also commend his eye for detail when examining various artefacts.

A mention must go to my close friend Rob Jelbart. Although he recently had to give up metal detecting due to serious health problems, he had the courage and tenacity to carry on searching fields with me until he was no longer able to do so. Rob has contributed numerous articles to *Treasure Hunting* magazine over the years and still has his heart firmly placed within the hobby. Mention also to Charles Smith who has sent me numerous images of his vast collection of finds.

Big thanks go to all fellow detectorists who have sent me drawings and photographs of their finds. In many cases they also included information on where the artefacts were recovered and other items found in local context. I feel privileged that they have given me their trust and they can be assured that this will be respected to the full.

For the excellent photographs I would like to thank Greg Payne whose professionalism does credit to this book.

Finally, I would like to thank my wife Brenda who has always encouraged me to continue detecting and writing, and also proof reads my articles and books. She accompanies me on all the talks I give both inside and outside the hobby, and is always willing to help.

Special mention must also go to:-
Alan Cunningham
Hubert Dabbers
Jon Alderson
Kevin Nicholson

LINE DRAWINGS

All line drawings are "life size" (also sometimes described as "To Scale", "Same Size" or "Scale 100%") in this or other publications unless otherwise stated.

PHOTOGRAPHIC ILLUSTRATIONS

Most photographs are 150% of life size unless otherwise stated. If a "cm" scale is shown this represents 1 centimetre alongside the object photographed and the actual size can be worked out in proportion (i.e. if the "cm" scale measures 2 centimetres then the object is 50% or half the actual size of the illustrated photograph).

Pendants

Pendants are as popular today as they were centuries ago, with some being made from base others from more precious metals. They are worn in most cases as decorative jewellery. However, there are still people who regard them as amulets, should this be for religious reasons or just for luck.

Our ancestors were little different in their thoughts and feelings than us. However, their belief in good fortune seemed to be a great deal stronger than ours and played a major part in everyday life.

Even before the Roman Occupation pendants were popular with the inhabitants of Britain including such designs as: wheels, daggers, effigies, axe heads, razors etc.

The Romans were also fond of pendants and produced and wore them in numerous shapes and designs, including some already favoured by the Celts.

One of the most common designs of pendants at this period was the phallic – not only in this country but

Fig.1. Roman bronze dagger pendant with hole in the blade used as a suspension loop; these were favoured by gladiators.

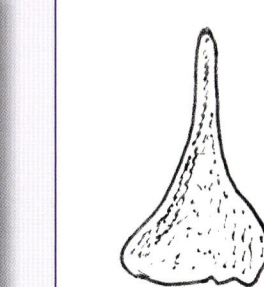

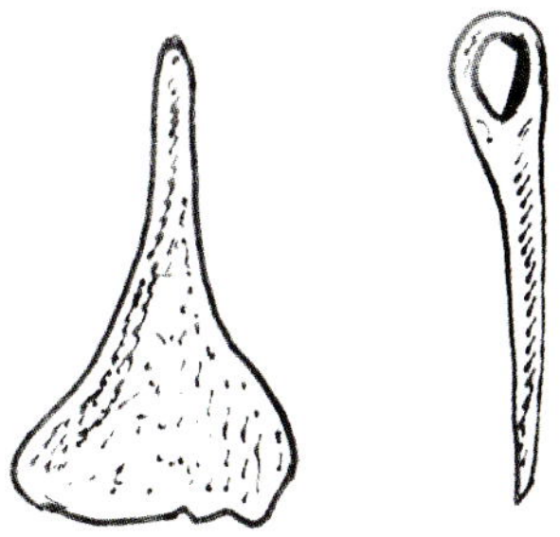

Figs.2a & b. Roman bronze pendant. This could be either an axe or razor but more likely the latter.

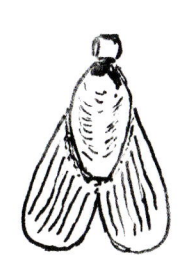

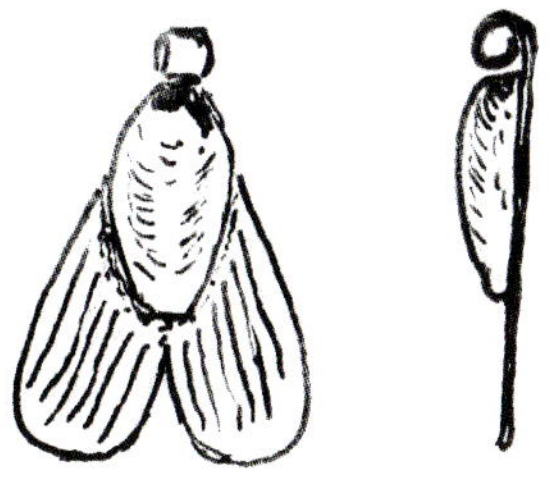

Figs.3a & b. Roman solid silver fly pendant.

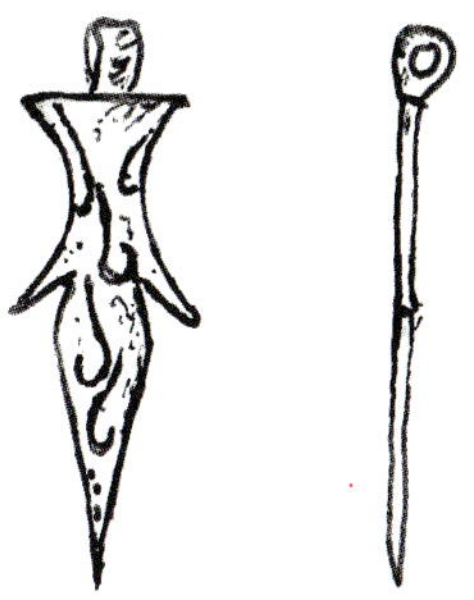

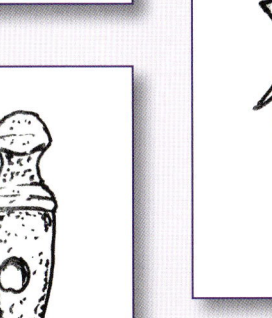

Figs.4a & b. Roman solid silver pendant engraved with swirls; these were favoured by the military and gladiators.

Fig.5. Roman bronze pendant in the form of a domestic knife.

Fig.6. Roman bronze pendant in the form of a sword or dagger; the blade is holed for use as a suspension loop.

Fig.7. Roman bronze axe pendant with suspension loop.

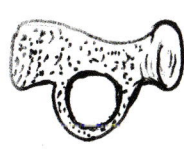

Fig.8. Roman bronze axe pendant with large suspension loop.

5

Pendants

throughout Europe. In the main these were formed in bronze, although a few examples are known and have been found in silver and gold.

Phallic representations were thought to ward off the "evil eye", and many have been recovered from former areas of military occupation.

One known silver pendant is in the shape of a gladiator's sword, so possibly it belonged to such a person as a good luck piece. As to other designs it is only possible to surmise. Axe pendants could, for example, have been favoured by carpenters but – if such was the case – this trade must have been very widespread. It is more likely that miniature axes were the symbol of some god. Sword and dagger pendants were probably associated with the military.

Crescent pendants also seem to have been a great favourite with some being highly decorated and others

Figs.9a & b. Roman bronze pendant; this is a smaller example of Fig.2.

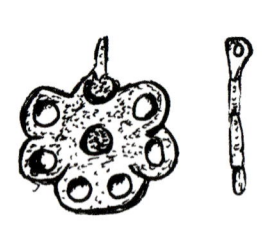

Figs.10a & b. Roman bronze pendant, petal-shaped with small loop.

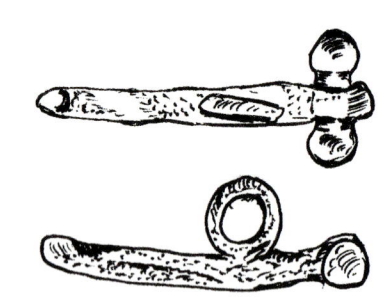

Figs.11a & b. Roman bronze phallic pedant. These were widely favoured and used throughout the Roman Empire.

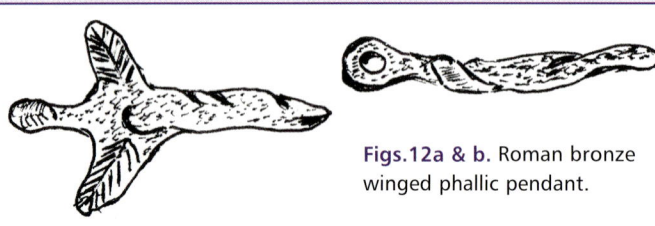

Figs.12a & b. Roman bronze winged phallic pendant.

Figs.13a & b. Roman bronze phallic pendant.

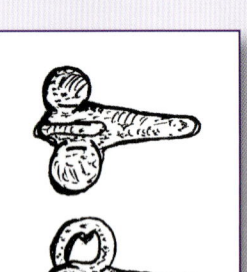

Figs.14a & b. Roman bronze phallic pendant.

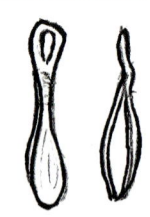

Figs.15a & b. Roman silver pendant in the form of tweezers. These have also turned up on Roman votive sites.

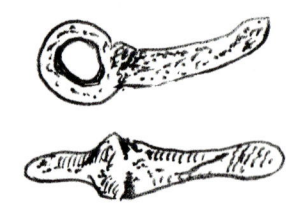

Figs.17a & b. Roman bronze phallic pendant. The suspension loop is rather crudely finished.

Fig.16. Roman bronze oval-shaped pendant terminating in a V-shape.

Pendants

quite plain. Bronze examples, in many cases, carry more decoration than those made of precious metal.

One of the most unusual pendants is the shield or boss type. There is also a Roman silver pendant in the form of tweezers, although who would have worn these is uncertain.

In the 5th century the Saxons replaced Roman rule, and the new settlers brought with them their own culture, and design of artefacts. However, many of the coins that the Romans left behind (whether still circulating or found on fields or abandoned settlements) were made into pendants or necklaces. Bronze Roman coins have been found in Saxon contexts with single or multiple holes depending on their intended decorative use. Silver Roman coins were used usually as a single piece, and gold coins must have been much sought after to convert to jewellery.

Figs.18a & b. Roman bronze phallic pedant with large suspension loop.

Fig.19. Celtic silver finger ring, which has a small axe head attached also made from silver. It is unlikely that it would have been worn as finger ring, more probably as a pendant.

Fig.20. Small Roman coins are found with purpose-made holes, varying between one and six in number. It is likely that these were worn as pendants, necklaces or some other form of decoration.

Fig.21. Late Celtic bronze pendant in the shape of a razor. Som times these have been described as "votive".

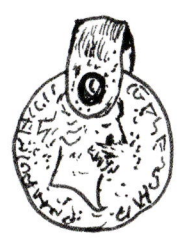

Fig.22. Roman silver coin made into a pendant during the Saxon period. The suspension loop (as is the case with the rivet) are of copper alloy.

Fig.23. Roman silver pendant in the shape of an amphora.

Fig.24. Roman silver pendant in the shape of a jug.

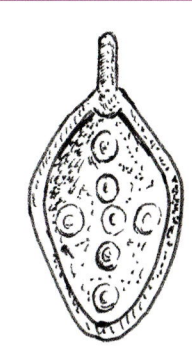

Fig.25. Saxon silver oval-shaped pendant with an outer double lined border and ring and dot design.

Pendants

Roman gold coins have been recovered with Saxon gold loops attached, and others enclosed within a decorated gold mount; these would have been worn singularly as a pendant or in multiples as a necklace.

At the other end of the scale the Saxons also favoured lead jewellery as is known from some of the items recovered. Saxon lead pendants in the record have somewhat crude figures and designs.

Many of the known Saxon pendants – whether of lead, bronze, silver or gold – have been recovered from graves. A bronze example from Finglesham is in the form of a head with a horned headdress with birds.

Some silver pendants were plain, or decorated with a ring and dot design (or other designs), some having three or four holes. Gold pendants were normally circular, although other shapes were favoured, and generally highly

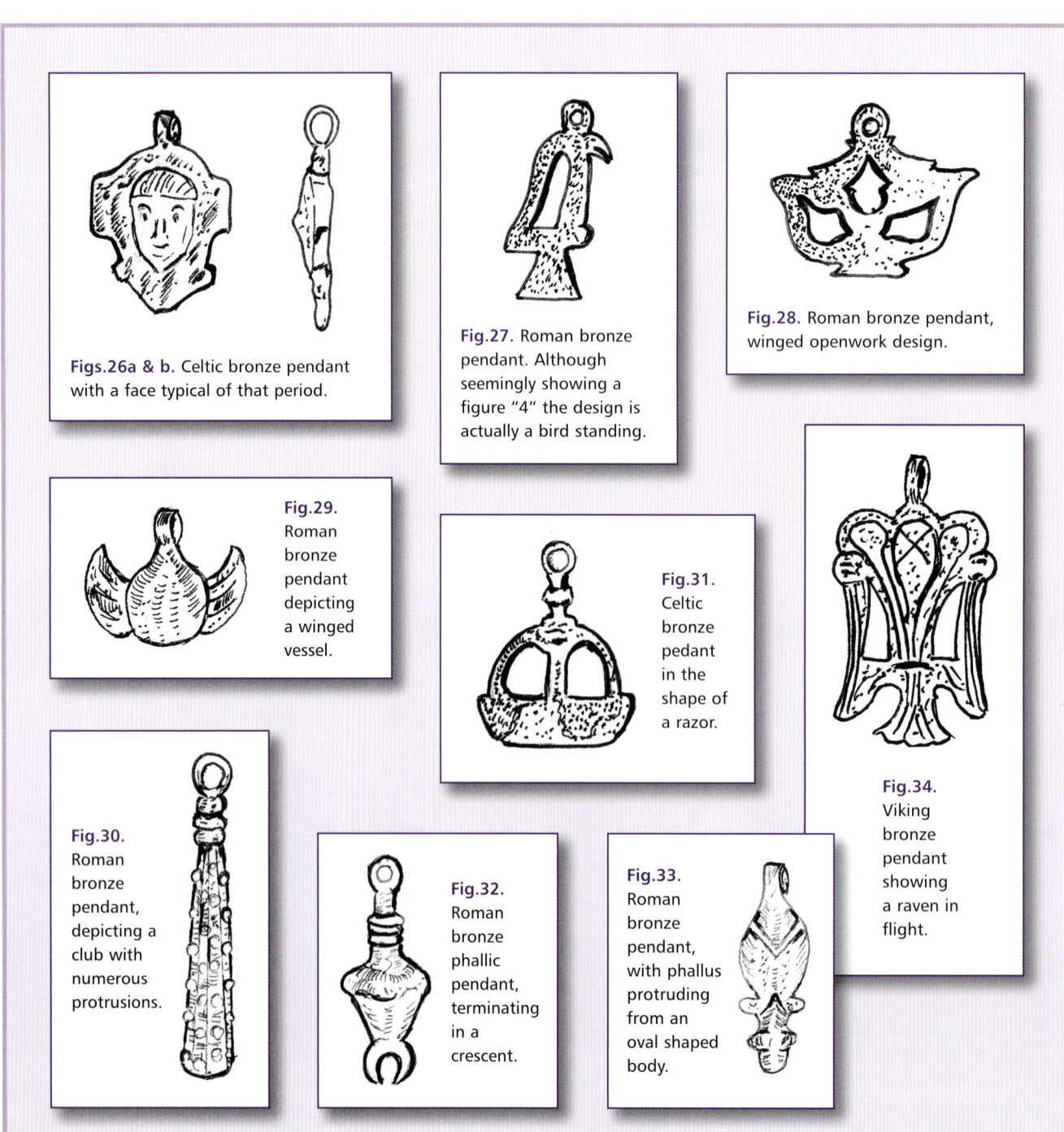

Figs.26a & b. Celtic bronze pendant with a face typical of that period.

Fig.27. Roman bronze pendant. Although seemingly showing a figure "4" the design is actually a bird standing.

Fig.28. Roman bronze pendant, winged openwork design.

Fig.29. Roman bronze pendant depicting a winged vessel.

Fig.31. Celtic bronze pedant in the shape of a razor.

Fig.34. Viking bronze pendant showing a raven in flight.

Fig.30. Roman bronze pendant, depicting a club with numerous protrusions.

Fig.32. Roman bronze phallic pendant, terminating in a crescent.

Fig.33. Roman bronze pendant, with phallus protruding from an oval shaped body.

Pendants

decorated. Some were adorned with a single central stone such as garnet, while others had a central garnet surrounded by other small garnets.

Towards the end of the first millennium the shores of this land was invaded once again; this time it was the turn of the Vikings. These warriors came from Norway, Sweden and Denmark, and for some 300 years from the 8th to 11th centuries terrorised the inhabitants of much of Europe, taking not only their gold and silver but also the people as slaves.

However, the Vikings were not only raiders whose name made people tremble with fear, but they were also traders, excellent navigators, craftsmen and ship builders. It was as the latter that allowed them to conduct their raids.

They believed in many different gods and goddess each having their own form of power. The main gods

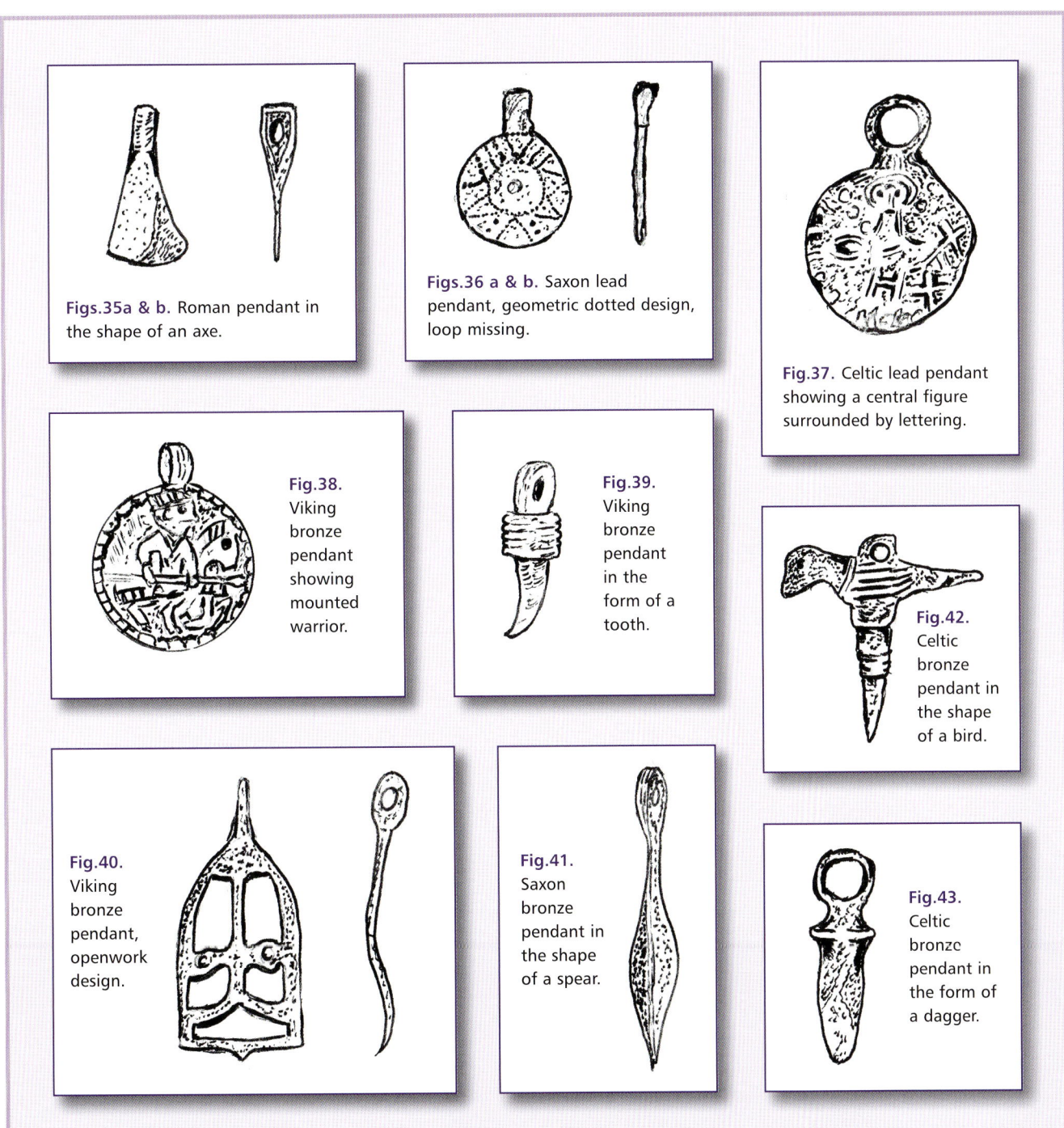

Figs.35a & b. Roman pendant in the shape of an axe.

Figs.36 a & b. Saxon lead pendant, geometric dotted design, loop missing.

Fig.37. Celtic lead pendant showing a central figure surrounded by lettering.

Fig.38. Viking bronze pendant showing mounted warrior.

Fig.39. Viking bronze pendant in the form of a tooth.

Fig.42. Celtic bronze pendant in the shape of a bird.

Fig.40. Viking bronze pendant, openwork design.

Fig.41. Saxon bronze pendant in the shape of a spear.

Fig.43. Celtic bronze pendant in the form of a dagger.

9

Pendants

were: Thor (who though not clever was extremely strong), Odin (who was the god of wisdom and war, and possessed many supernatural powers) and Frey (the god of fertility).

The pendants the Vikings wore were of base or precious metal, and created in numerous designs. The hammer of Thor is the most recognisable, for this god was popular with peasants and farmers. It was believed that he rode through the sky in a chariot that was pulled by goats. He used the hammer to fight evil giants and monsters, clubbing them to death.

Snake pendants were also favoured and worn as good luck charms. Some pendants have scenes such as a mounted warrior with a spear. Others take the design of a wolf's head, a tooth (possibly that of a bear), a spear, axe head (some plain others decorated), the head of Odin etc.

Coins were also used, normally

Fig.44. Celtic bronze pendant, figure of eight shape, depicting a typical Celtic face surrounded by a beaded border.

Fig.45. Viking bronze pendant in the shape of a decorated axe head.

Fig.46. Viking bronze pendant in the shape of a plain axe head.

Fig.47. Viking bronze pendant depicting a decorated wolf's head.

Fig.48. Viking bronze pendant showing Odin's head.

Fig.49. Viking highly decorated silver pendant depicting Thor's hammer.

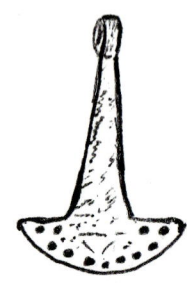

Fig.50. Viking silver pendant depicting Thor's hammer, the blade section having a dotted border.

Fig.51. Viking silver pendant depicting Thor's hammer. This is a small example, decorated with dots on the blade and handle.

Pendants

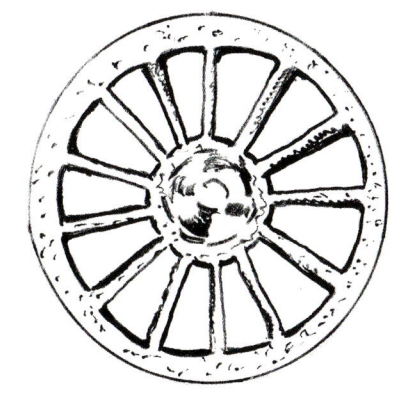

Fig.52. Bronze multi-spoke wheel pendant. These have been recovered on Bronze Age and Iron Age sites; but also on some Roman sites.

Fig.53. Bronze six-spoke wheel pendant. Examples of this type have mainly been recovered on Bronze Age sites.

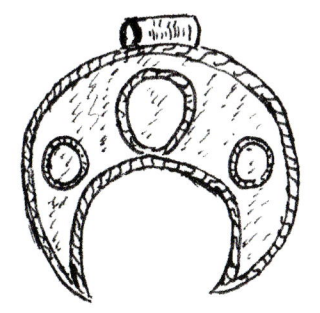 

Figs.54a & b. Roman gold crescent pendant. Rope design on outer edges with three roped circles and oblong suspension loop.

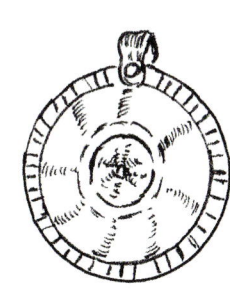 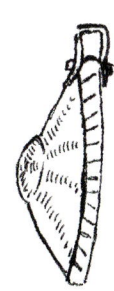

Figs.55a & b. Roman gold boss-shaped pendant, lined border, suspension loop attached by a gold rivet.

Fig.56. Roman silver pendant, heart-shaped with lower spherical design.

Fig.57. Roman silver crescent-shaped pendant, dot border.

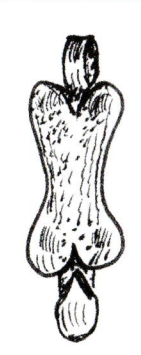

Fig.58. Roman bronze phallic pendant.

Fig.59. Roman bronze pendant, possibly depicting a heart and phallus.

Pendants

having a small suspension loop attached. Other items that individual Vikings acquired during a raid, and that took their liking (such as a belt decoration made from precious metal) would have been holed and worn as a pendant.

There are also odd pendants about which little is known. One in particular takes the form of the head of a Viking warrior showing much detail.

Viking pendants of gold with filigree and granulation would rival any

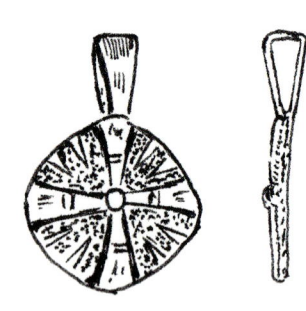

Figs.60a & b. Roman bronze circular pendant with equal-armed cross, and small central boss with pointed designs in each quarter.

Pic.1. Roman bronze dagger pendant with hole in blade used for suspension. This design was favoured by gladiators.

(Pics.1-8 are not to scale)

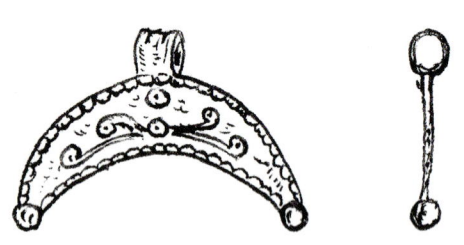

Figs.61a & b. Roman gold pendant, crescent-shaped, beaded border, central floral design.

Figs.63a & b. Roman bronze pendant, winged design, raised boss at top.

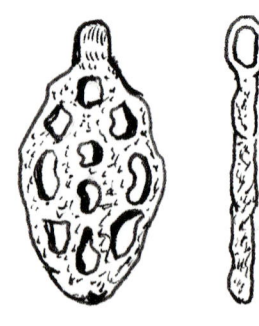

Figs.62a & b. Saxon silver pendant, irregular openwork design, and somewhat crudely made.

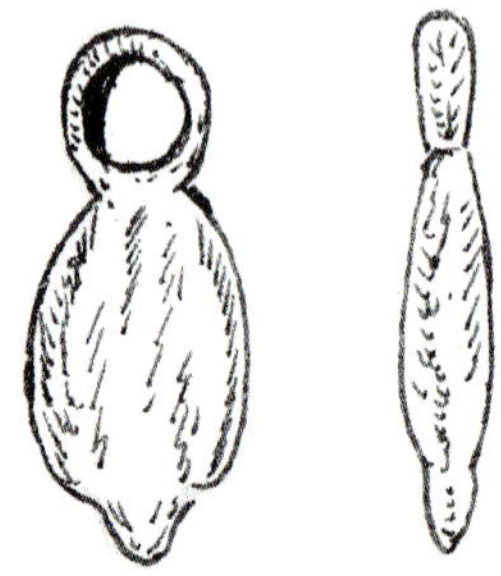

Figs.64a & b. Roman bronze pendant, oblong teardrop shape.

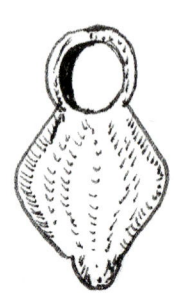

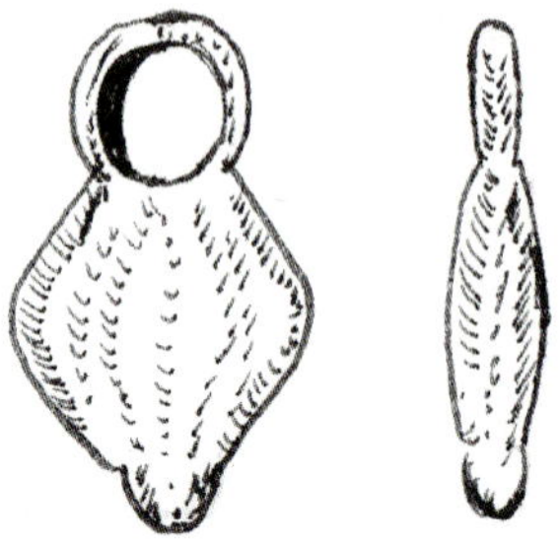

Figs.65a & b. Roman bronze pendant similar to Fig.62. but wider.

Pendants

Pic.2. Roman petal-shaped bronze pendant.

Pic.9. Roman bronze pendant, winged design, raised boss at top.

Pic.10. Roman bronze pendant, winged openwork design.

Pic.3. Roman bronze axe or razor pendant (probably the latter).

Pic.4. Roman bronze phallic pendant, terminating in a large loop.

Pic.5. Roman bronze phallic pendant with large ornate suspension loop.

Pic.11. Viking bronze openwork pendant.

Pic.12. Roman bronze pendant. Although the design looks like a figure "4" it actually represents a standing bird. *(Not to scale)*

Pic.6. Roman bronze phallic pedant with top suspension loop.

Pic.7. Roman bronze splayed phallic pendant with central top loop.

Pic.8. Roman bronze splayed phallic pendant, terminating in a large loop.

Pic.13. Roman solid silver fly pendant.

Pic.14. Saxon pendant made of lead; geometric dotted design, loop missing.

13

Pendants

modern jewellery, as would the silver examples they produced. Eventually the Vikings gave up their gods for Christianity and took to the wearing of a cross. However, it is likely that they hedged their bets for a while by wearing the cross along with Thor's hammer to be on the safe side. A stone mould recovered from Denmark has both the cross and Thor's hammer on it, showing that the craftsmen were making them both at the same time.

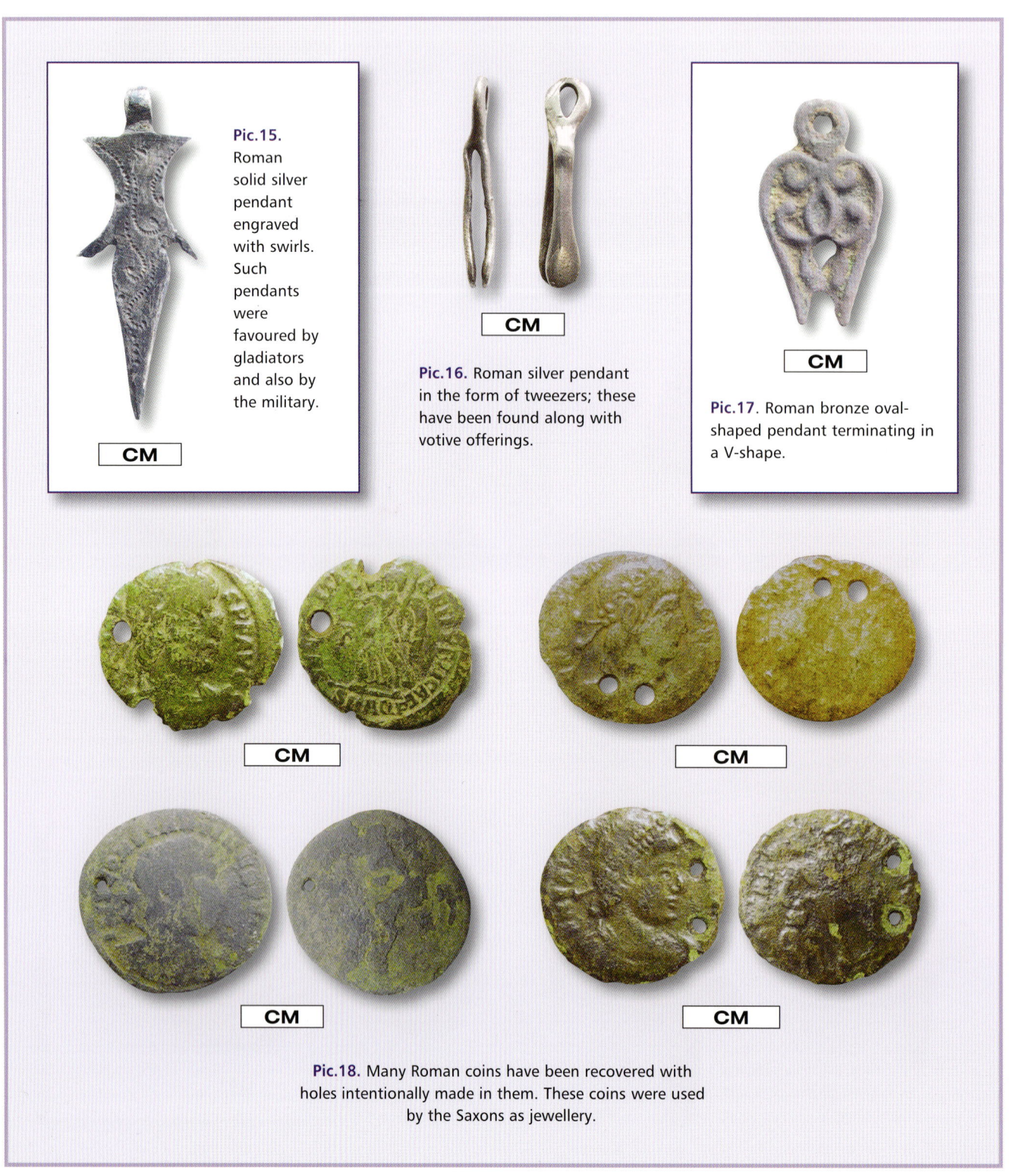

Pic.15. Roman solid silver pendant engraved with swirls. Such pendants were favoured by gladiators and also by the military.

Pic.16. Roman silver pendant in the form of tweezers; these have been found along with votive offerings.

Pic.17. Roman bronze oval-shaped pendant terminating in a V-shape.

Pic.18. Many Roman coins have been recovered with holes intentionally made in them. These coins were used by the Saxons as jewellery.

Open-Type Bells

Prior to the Roman invasion bells existed and are known to have been used in Britain. This use, however, became more widespread after the Occupation due to an increase in the functions that the Romans found for them. Besides decorative and practical uses, bells also had a role to play in ritual and religion.

Bells were fitted to animals for the dual purpose of warding off the "evil eye" and locating them when rounding them up after they had been released for grazing.

It is believed that the Romans hung bells around the necks of criminals when taking them to be executed. The practice was thought to ward off any evil emanating from the convicted person, and also serve as a warning to others who might contemplate breaking the law.

In some European countries, even at a much later date, criminals taken to be executed were fitted with bells around their necks. Once again, this was probably as a warning to others not to break the law.

The main use of Roman bells was more positive, however, and they were widely used in celebrations such as triumphs, dances and festivals. Besides the use of the open-mouth bell at these occasions, an early form of the crotal bell was employed along with other instruments.

The Romans also opened and closed the Games with a bell, as they did for Imperial Triumphs. In fact, it is surprising how much an important part bells played in Roman life. The Emperor Augustus had a dream in which he was warned that the great doors of the Temple of Jupiter in Rome should have bells fitted to them. Roman sentries would carry bells both to announce the hour and also to warn of any potential attack (the noise made by a bell would

Fig.1. Roman bronze beehive-shaped bell with pyramid loop, plain body terminating in a raised rim.

Fig.2. Roman bronze bell, pyramid loop terminating in a splayed base, plain body.

Fig.3. Roman bronze bell small round loop, terminating in wide splayed plain body

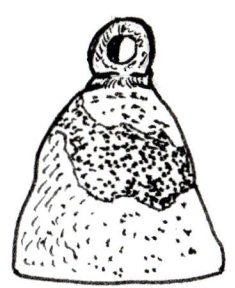

Fig.4. Roman bronze bell of beehive form, crudely-shaped thick round loop, and plain body. Similar types were also used in the Saxon period.

Fig.5. Roman bronze bell, large pyramid-shaped loop, cup form body having a central double line around the perimeter.

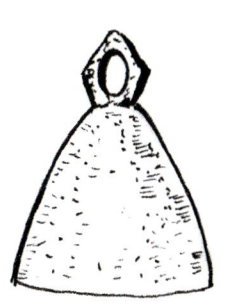

Fig.6. Roman bronze bell of beehive shape, pyramid type loop, plain body.

Fig.7. Roman bronze bell, pyramid-shaped loop and pyramid form plain body terminating in four flat feet (possibly for votive use on altar).

Open-Type Bells

Fig.8. Roman bronze bell, five-sided, loop of beehive shape, splayed base with a lined border terminating in four flat feet (possibly for votive use on altar).

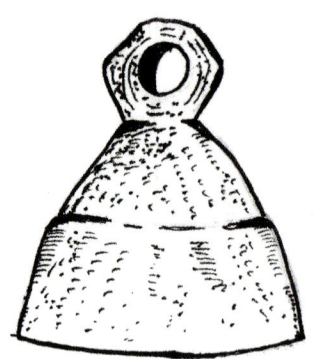

Fig.9. Roman bronze bell, beehive shape, large five-sided loop, plain body with raised central rim.

Fig.10. Roman bronze bell, five-sided loop, central raised line repeated at lower part of body terminating in a lined border (*tintinnabula* type).

Fig.11. Roman bronze bell of beehive shape, crude round loop, plain body (*tintinnabula* type).

Fig.14. Roman cup-haped bronze bell with very small loop and two double lines around centre. These were used on arm and ankle bracelets and possibly associated with dancing.

Fig.12. Roman bronze bell of beehive shape, crude oval loop, raised double banded border around centre (*tintinnabula* type).

Fig.13. Roman bronze bell of beehive shape, five-sided loop, double line around the centre (*tintinnabula* type).

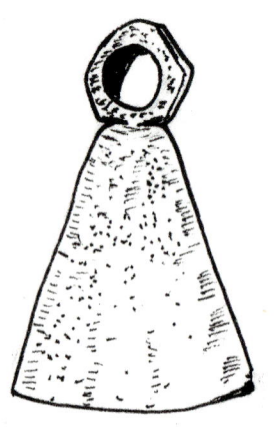

Fig.15. Roman bronze bell of slender form, large pyramid loop, plain body.

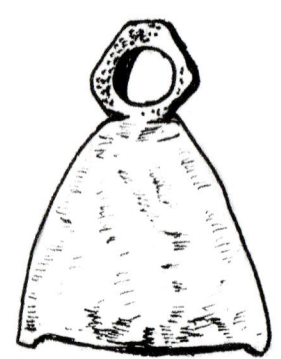

Fig.16. Roman bronze bell of beehive shape, five-sided loop, the body terminates in four small feet (possibly for votive use on altar).

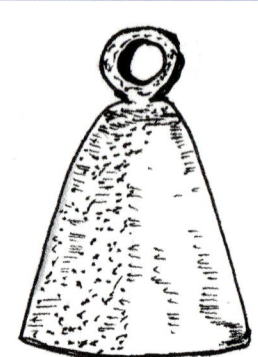

Fig.17. Roman bronze bell of slender form, round loop, plain body.

Open-Type Bells

carry much further than the noise of a person shouting).

Small bells were used inside Roman temples by priests and initiates although little is known about their ceremonial use. In Rome itself, wives suspected of being unfaithful were forced to wear bells and ride a donkey through the streets as a form of public spectacle.

Cleanliness for the Romans was also a leisure pursuit. At around midday a bell was rung to let the local inhabitants know that the water was warm and that the baths were open. Here they could meet friends and discuss the topics of the day.

An array of five small bells – each suspended separately from one main holder – was hung in shops and doorways to keep out evil spirits and protect those within. Such an arrangement was known as a *tintinnabulum* (wind bells).

Larger bells were also hung outside the homes of some people. One Roman example from the 2nd or 3rd centuries that was found in Switzerland was 16.7cm in diameter and 10.2cm high.

Bells of this or similar size would have had a dome-shaped loop, for suspension from a chain or cord. The iron clapper, that was longer than the height of the bell, would have terminated in a loop. To this a leather strap or cord was attached and the bell would have been rung in the same fashion as a ship's bell.

Some Romans also adorned their

Fig.18. Roman bronze bell, beehive-shaped with round loop pierced with very small hole; openwork Vs around the whole body.

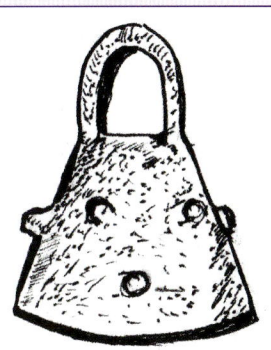

Fig.19. Roman bronze bell of slender form but with very large loop; the body has numerous protruding lugs.

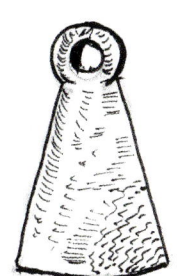

Fig.20. Roman bronze bell of slender form with round loop and plain body.

Fig.21. Roman bronze bell the whole terminating in four small feet. This example is very crudely made and was possibly intended as a votive offering.

Fig.22. Roman bronze bell with five-sided loop that is still attached to some of its original chain; it has two double lines around the top and lower half. This would have been too large for a *tintinnabulum*, but was possibly hung from a doorway.

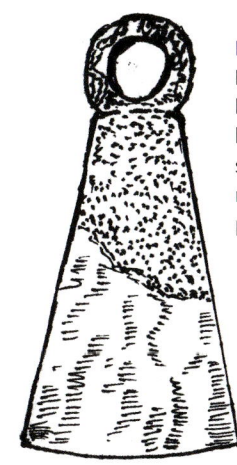

Fig.23. Roman bronze bell of long slender shape, large round loop, plain body.

Fig.24. Roman bronze bell, beehive shape, round loop, plain body.

Open-Type Bells

Fig.25. Roman bronze bell, cup shaped with five-sided loop, double line around base, possibly a *tintinnabula*.

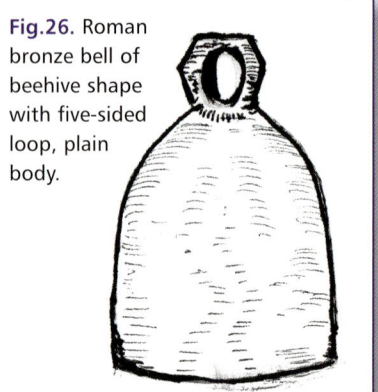

Fig.26. Roman bronze bell of beehive shape with five-sided loop, plain body.

Fig.27. Roman bronze bell of beehive shape with crude round loop, the whole terminating in a single lined plain body (possibly made as a votive offering).

Fig.29. Roman bronze bell of slender shape with large five-sided loop, plain body.

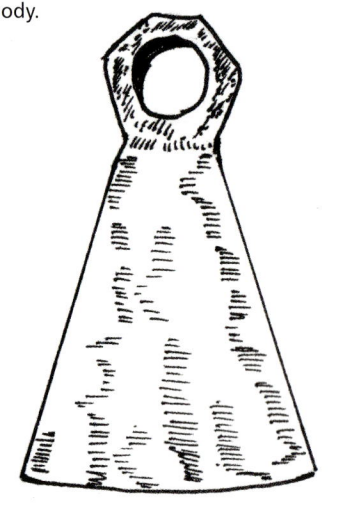

Fig.28. Roman bronze bell of cup shape, oval loop, single line around base with a scratched inscription above. Such large bells could have been outside dwellings or even temples, and used in the same manner as a modern-day door bell.

Fig.30. Roman bronze bell (similar in shape to Fig.27.) with two double lines at top and base, but no inscription.

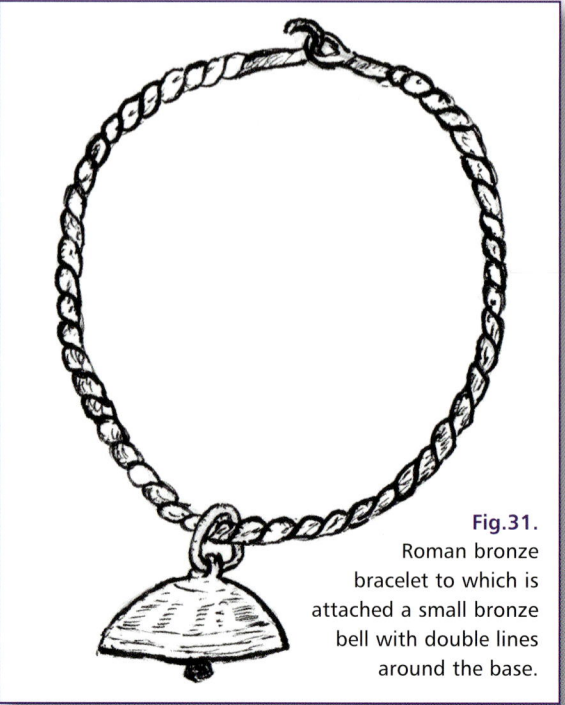

Fig.31. Roman bronze bracelet to which is attached a small bronze bell with double lines around the base.

Open-Type Bells

horses with small bells, some examples of which have been found near Hadrian's Wall. Such were normally small and made from bronze although the clappers were usually constructed from iron.

Small bronze bells of lightweight construction were also worn by the Romans as items of personal decoration. An example of one of these was recovered during an excavation at Colchester in Essex and dates early 4th to mid 5th century. This small bell was attached to a twisted bronze wire bracelet along with a small ornate bead.

These small bells are similar to the *tintinnabulum* in style but the metal is thinner. The loop is also much smaller, with a hole just large enough to allow a thin wire to be passed through.

Religious or votive bells of Roman date have also been recovered, some examples having been fitted with small feet. These would date prior to the early 5th century for in 408 Theodosius issued an edict forbidding all forms of non-Christian worship throughout the Roman Empire. This would have included the use of such pagan bells.

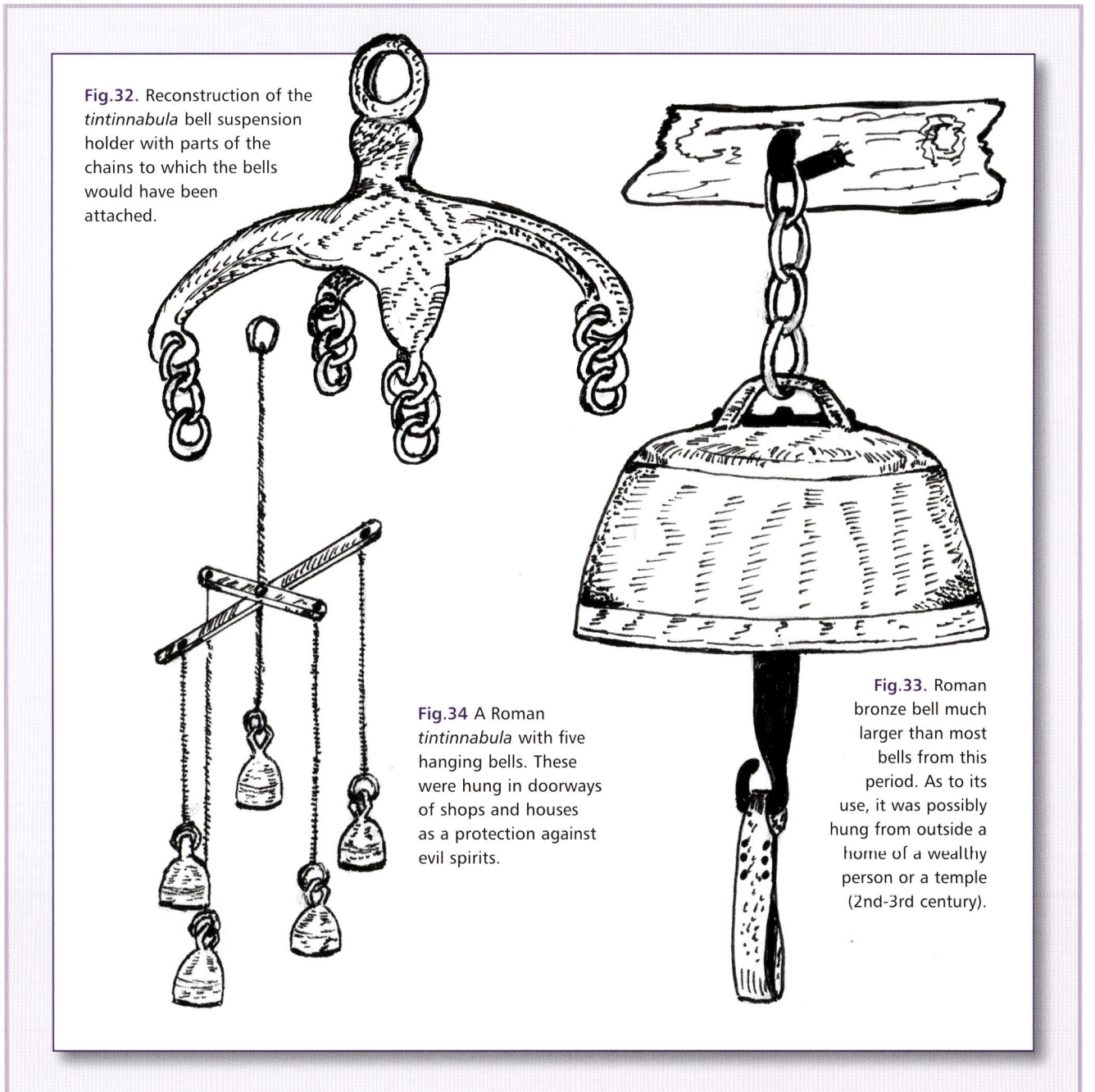

Fig.32. Reconstruction of the *tintinnabula* bell suspension holder with parts of the chains to which the bells would have been attached.

Fig.34 A Roman *tintinnabula* with five hanging bells. These were hung in doorways of shops and houses as a protection against evil spirits.

Fig.33. Roman bronze bell much larger than most bells from this period. As to its use, it was possibly hung from outside a home of a wealthy person or a temple (2nd-3rd century).

Open-Type Bells

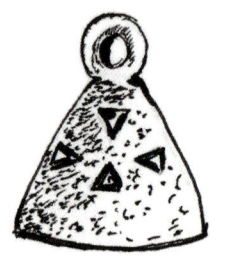

Fig.35. Saxon bronze bell of beehive shape with round loop and small pyramid cut-outs.

Fig.36. Saxon bronze bell of six-sided form, each segment divided by raised edges terminating in a foot. Each panel is decorated with four rings and dots. The bell has a three-sided loop above a collar.

Fig.37. Saxon bronze bell, plain body, small round loop.

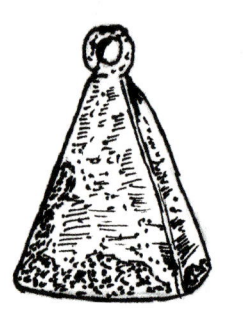

Fig.38. Saxon bronze bell four-sided plain body, small round loop.

Fig.39. Saxon bronze bell, four sides, each with raised edges; plain body, large three-sided loop above a collar.

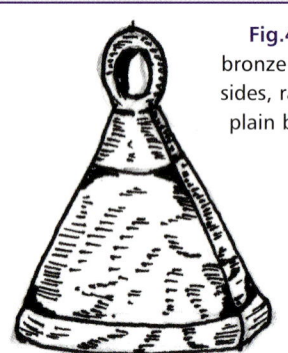

Fig.40. Saxon bronze bell, four sides, raised rim, plain body, oval loop.

Fig.41. Saxon bronze bell, beehive shape, plain body, large round loop.

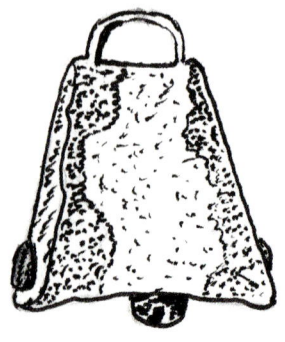

Fig.42. This style of iron sheet bell with side iron rivets and iron clapper was used from the Saxon period until the early 20th century. The style and methods of manufacture remained unchanged throughout all those years.

Fig.43. Small Saxon bronze bell, plain body, large oval loop. Such examples would have been used as pendants worn around the neck on a chain or leather thong, or on a wrist bracelet. Such bells were used by the wearer to ward off evil spirits.

Open-Type Bells

Saxon & Viking Bells

When the Romans left our shores to protect their remaining Empire, it was the turn of the Saxons, Angles and Jutes to settle here, and eventually the Vikings. During an archaeological excavation at Flixborough a lead tank was recovered inside of which was a set of carpentry tools (including axes, adzes, and spoon bores) together with a bell.

The latter was of an open type being made from a sheet of iron about 2mm thick, that had been folded into shape and then riveted. (This type of bell is now known as a "clucket", although in the Saxon period it would have been known by another name). The clapper was also being made from iron and terminated in a bulbous head.

The handle (missing) would have been in the form of an elongated "M" or else "D" shaped. It is likely that it was the latter, as there is a fragment of a Saxon cross showing a bell ringer holding two bells; one of these depicts the "D" shaped handle. The ringer is using the bells in the drop arm method. The style of these bells would indicate that they are of the clucket type as the handles are large enough for the fingers to pass through.

Larger examples of these iron bells were reputed to have been used by those who went about preaching the word of God; one such example was said to have been used by St. Conall. One of the most famous of these bells is the that of St. Patrick Will or

Fig.44. Pewter bell with inscription around the upper part reading "CAMPANA THOME" within a double line, and trefoil loop. These are often called "Canterbury bells" (14th-15th century).

Fig.45. Pewter bell with an inscription around the base reads" CAMPAN THOME A" and trefoil handle. These are, again, often called "Canterbury bells".

Fig.46. Pewter bell of dome shape, decorated rim in the form of zigzags below a double line, oval loop, 14th-15th century. These are, again, often called "Canterbury bells".

Fig.47 Pewter bell consisting of four crescent shaped segments each showing the letter "S"; small round loop.

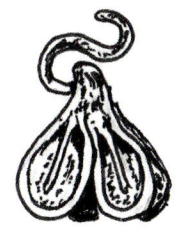

Fig.48. Pewter bell consisting of five decorated panels petals each having a lined design. The "S" shaped loop passes through the top of the bell, 14th-15th century. These are, again, often called "Canterbury bells".

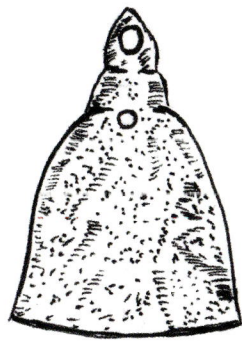

Fig.49. Bronze bell, plain body, pyramid-shaped loop below which can be seen a small hole. This is where the rivet bar would have been that was used for suspending the clapper (14th-16th century).

21

Open-Type Bells

Clog-a-eadbacta Phatraic. This bell is believed to be the one carried by St. Patrick into Ireland in the 5th century, and is said to be have been used by him when performing miracles. This bell is now housed in a gold and jewelled shrine and can be seen in The National Museum of Ireland in Dublin. Religion played a great part in the lives of our ancestors. The Irish disciples took the early form of Christianity to Iceland, and then over the Continent from the Baltic to the Mediterranean. The tone of these bells must have been quite dull, but they served their purpose.

Besides iron, some bells of this type were made from hammered copper. A number of such bells of this period exist in France, Germany, Switzerland, Scotland and Ireland.

It is interesting to note that the name bell is from the Saxon word "bellan" to roar; also the Old Norse "bjalla" (from "bylja") to resound. A hollow cup shaped receptacle normally made from metal, and when struck on the rim, would make a sound.

Surviving small bronze bells from the Saxon and Viking periods are few in number and little is known of their intended uses. Some examples are very plain while others highly decorated, which could indicate that they were intended for ceremonial use.

It is difficult to distinguish between bells used by the Vikings and those by the Saxons, as there was much trading going on between the two groups.

Fig.50. Bronze bell of beehive shape. It carries the design of the face of man, looking somewhat unhappy, on both sides. Oval-shaped loop, 14th-16th century.

Fig.51. Pewter bell with plain body, 14th-15th century. The casting line is prominent along the top of the loop and down both sides.

Fig.52. Pewter bell, with six sides splaying out to form petals of a flower, trefoil loop, 14th-15th century. These are often called "Canterbury bells".

Fig.53. Bronze bell, line decoration around the base, crudely-shaped loop, 14th-16th century.

Fig.54. Pewter bell with inscription around base too worn to decipher, oval-shaped loop, 14th-15th century. These are often called "Canterbury bells".

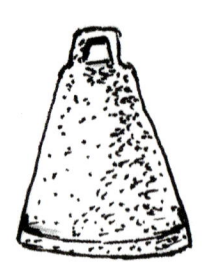

Fig.55. Bronze bell, line decoration around base, square-shaped loop, 14th-16th century.

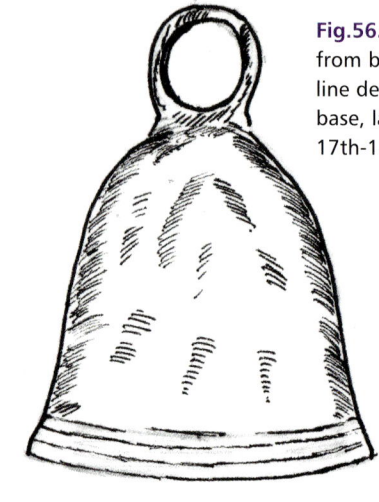

Fig.56. Horse bell, cast from bell metal, three line decoration around base, large round loop, 17th-18th century.

Open-Type Bells

Norman & Medieval Bells

A scene in the Bayeux Tapestry (woven about 1100) shows acolytes walking by the coffin of Edward the Confessor (who died 1066) on its way to interment in Westminster Abbey. The acolytes have two "spirit bells" that were believed to keep evil spirits away from the body of the deceased, allowing it to reach its final resting place unaffected and then placed within the holy building.

One of the most well-known bells must be the pewter Canterbury bell, which is often mentioned in manuscripts of the period and associated with pilgrims.

Such small open type bells used by the pilgrims were normally in the shape of church bells and date from the 14th century into the first half on 15th century. The early types of these miniature bells had an inscription around the lower part of the rim. This inscription became shorter and smaller over the years and was eventually replaced by a simple zigzag decoration or something similar.

The lower part of the bell at this time became splayed out in the form of an upside down fluted flower. Loops at this period were normally oval, trefoil or sometimes S-shaped.

One of the designs known for these small early medieval bells is that of the face of a man with very large eyes and a turned-down mouth. Whether intended or not, the depiction has the appearance of a death mask.

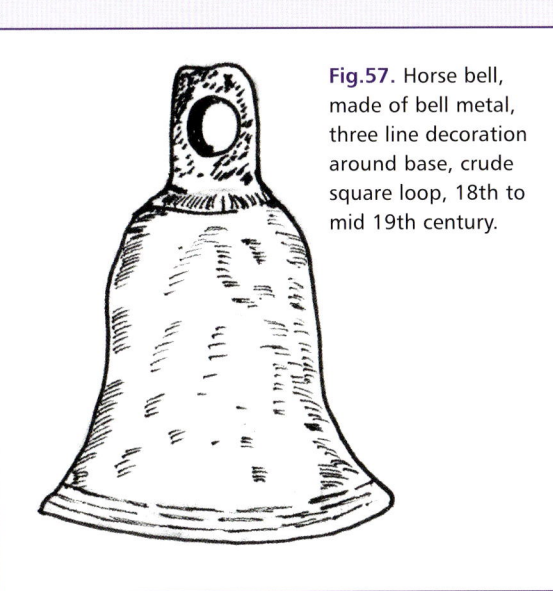

Fig.57. Horse bell, made of bell metal, three line decoration around base, crude square loop, 18th to mid 19th century.

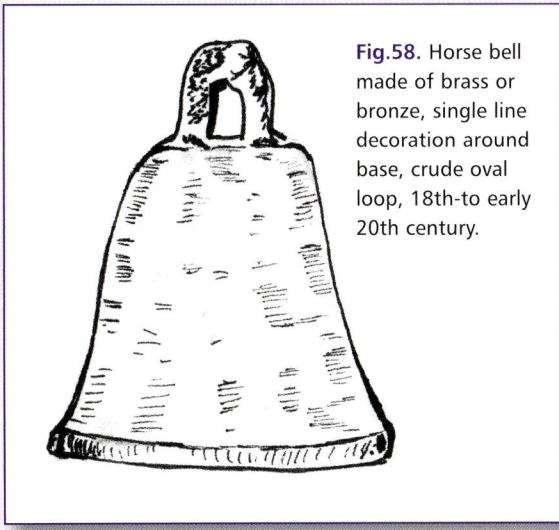

Fig.58. Horse bell made of brass or bronze, single line decoration around base, crude oval loop, 18th-to early 20th century.

Fig.59. Horse bell made of brass or bronze, two double line decoration around base, maker's mark showing a bell, small oval loop above a double line, late 18th to late 19th century.

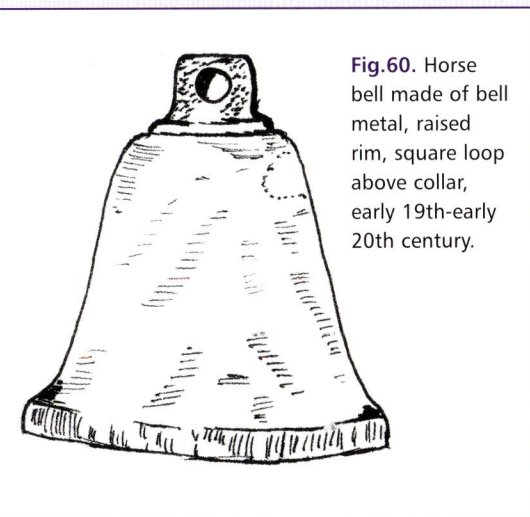

Fig.60. Horse bell made of bell metal, raised rim, square loop above collar, early 19th-early 20th century.

Open-Type Bells

The hand bell, being portable, played a great part in everyday life in medieval times. Unlike the static tower-housed bell, it could be taken to various events, such as fairs, markets, and social gatherings. In fact, certain villages and towns had their own events and customs in which the hand bell played an important part.

One common custom was performed by a minor church official known as a "beadsman" or "bedesman", for which he received a stipend. His duty was to walk around the parish ringing his bell and thereby announcing his presence. Having attracted the attention of the inhabitants he would offer prayers with those who had a death in the family, or who had an anniversary of a death falling at that time.

There was another bell associated with death, and which put fear into the hearts of the bravest souls in medieval England. This announced the approach of a leper. This terrible disease was widespread in England from the 11th to 13th centuries. A leper was not allowed to walk the streets unless he wore a cloak that reached down to his or her feet, and a hood that covered the head. In one hand a bell was carried that was constantly rung to give warning of his passage, and in the hope that some alms might be received.

The other bell associated with death

Fig.61. Horse bell made of bell metal, single line decoration around base, round loop on raised collar, 18th-early 19th century.

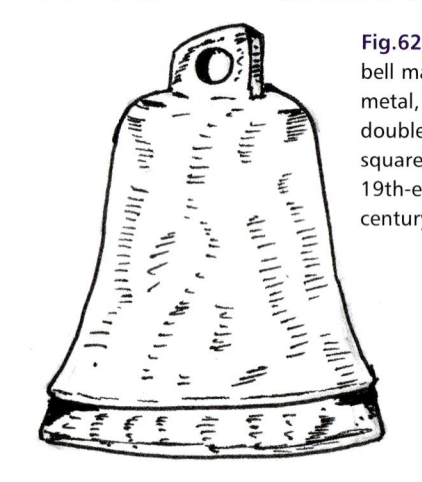

Fig.62. Horse bell made of bell metal, unusual double base, square loop, 19th-early 20th century.

Fig.63. Horse bell made of bell metal, raised rim above which double line decoration, square loop, early 19th-20th century.

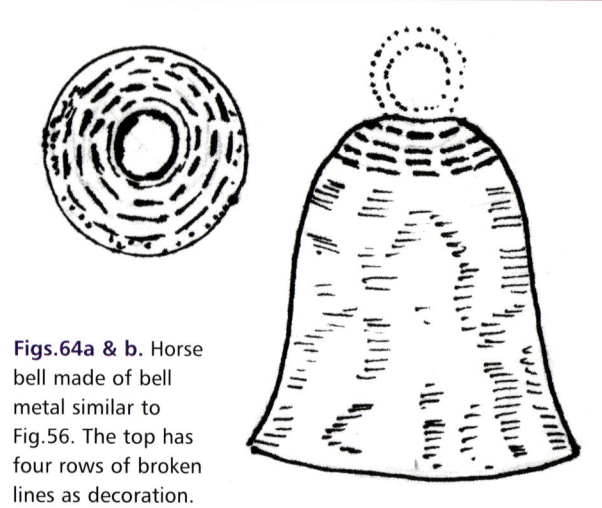

Figs.64a & b. Horse bell made of bell metal similar to Fig.56. The top has four rows of broken lines as decoration. The loop, although missing, would have been large and round, 17th-18th century.

Open-Type Bells

was used in the mid 1660s and in particular in 1665. This was at the time of the Great Plague, which in London alone claimed 70,000 people of all ages. Those who died from the plague were collected by carts and taken for burial. At first these operated only at night, but when the plague was at its height the carts were in use both by day and night. The men employed to collect the bodies would tour the streets ringing a hand bell and calling "Bring out your dead".

During the English Civil War not all the fighting involved major battles. Many minor skirmishes took place, some near a village or town. Some of these would have involved the inhabitants, and to give warning of an impending attack a hand bell was rung in an upright position giving full force to the noise so that it could be heard by all.

The watchmen who walked the streets of towns calling out the hours also used hand bells. Pepys wrote in his diary:

I staid up until the bellman came by and cried "Past one of the clock and a cold, frosty windy morning".

Many of these old customs have died out and no longer have any importance; however, there is one of these old customs that has made a revival in many towns – the town crier.

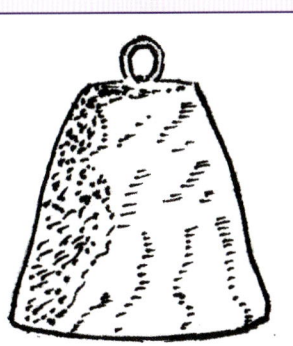

Fig.65. Bronze bell, possibly horse, plain body, small loop, 15th-17th century.

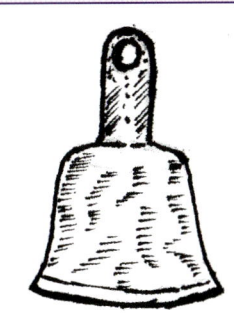

Fig.67. Horse bell made of bell metal, decorative line around rim, tall round top loop, 19th-early 20th century.

Pic.1. Horse bell, small oblong loop, late 18th to early 20th century.

Fig.66. Horse bell made of bell metal, single line decoration around base, tall loop, 19th-early 20th century.

Note: Figs.56-67. The bells shown, although originally made for the horse, were also used on sheep and other livestock.

Fig.68. Sets of bells were worn by horses in the wagon teams. This was to warn of their approach on the narrow roads and lanes where passing each other would have been difficult.

Open-Type Bells

Pic.2. Medieval bronze bell, plain body, pyramid-shaped loop, 14th-16th century.

Pic.4. Horse bell with raised rim and crude loop, 18th-early 20th century.

Pic.3. This style of iron sheet metal bell with side rivets and iron clapper was used from the Saxon period until the early 20th century. The style and indeed the manufacture remained unchanged throughout all those years.

Pic.5. Horse bell, unusual that it is made of bronze, oblong loop, ribbed top, 18th-early 19th century.

Open-Type Bells

Bells were fitted to many animals such as sheep and goats, but in particular horses. Before the internal combustion engine the horse played a great part in the everyday lives of many people. Besides ploughing they were used for transport be it for pulling carriages or coaches, or as saddle horses for individual riders. They were also harnessed to wagons for transporting food, clothing and other goods. Packhorses were also used to transport merchandise with the goods being strapped to their backs. In the latter case the horses were roped in single file and fitted with bells. This was so that should one become un-tethered and stray from the main group, it could easily be located by the sound of its bell or bells.

The other method of transporting goods was by horse and wagon, but this form of transport required a reasonable road surface on which to travel. Horses drawing wagons were also fitted with bells, but not in the same manner as the packhorses. Instead of being secured individually to the horse, bells were hung in sets from a purpose-made frame. This frame was referred to as a "box" or a "belfry". It was a small wooden beam that had iron supports on either side. These were shaped to fit into staples on the wooden hames. A sheet of leather was used to cover the beam; this was not only to protect the wood but also the bells from the weather.

Some of these belfries also had additional decoration such as small brass studs and brasses, woollen fringing, or scalloped leather. The bells that were attached to these boxes were not

Pic.6. Horse bell, three decorative lines around base, large round loop, 17th-18th century.

Pic.7. Horse bell made from bronze with small loop and splayed base. Napoleon's horse contingents are known to have used such bells on ceremonial occasions, 18th-19th century.

Pic.8. Roman bronze bell, two lines of decoration at top and base. Such large bells could have been fitted outside dwellings or even temples, and used in the same way as a modern day doorbell.

Pic.9. Iron bell made of sheet metal. This example is much larger than that shown in Pic.3., although its use was still the same.

Open-Type Bells

Pic.10. Left to right: Roman bell with line decoration around the rim, large loop. Part of a Roman *tintinnabula* type (broken). Part of a Roman bell, protruding lugs, large loop. Part of a Roman bell of beehive shape, crude loop. Centre part of an 18th century horse bell.

Pic.11. Roman bell of pyramid shape, plain body terminating in four flat feet, pyramid form loop, possibly for votive use on an altar.

Pic.12. Roman *tintinnabula* bell with line decoration around the top.

Pic.13. *Tintinnabula* suspension holder. Chains would have been attached to each of the four arms from which small bells were hung.

Open-Type Bells

only of the open-mouth type, but were often mixed (such as one open type with a rumbler on either side). Some boxes would have had only rumbler bells while others different sized open-mouthed types.

One bell founder in particular was well known for casting both rumbler and open-mouthed bells, this being Robert Wells of Albourne in Wiltshire.

The box or belfry was popular in the southern part of England. However, in Herefordshire large iron loops covered in leather were used to support the bells above the hames and collars, and a special strap was used to steady them to the saddle. Some of these loops housed six, eight or 10 bells of various size. The largest of these loops measured 3 feet in width and stood 2 feet above the fittings. The poor horse must have found the additional weight quite a burden.

An unusual yet practical use of open type bells, still employed in some places today, is to give warning of danger such as causing damage to overhead train cables. Rows of bells are strung above a level train crossing at even intervals of one metre, the whole being attached to a thick wire. Should any vehicle attempt to cross the railway line that is too high, then it will come in contact with the bells causing them to ring. One might think that in this day and age an electronic system would be more appropriate; however, there is less to go wrong with the old method.

Until recent times many small shops had an open type bell attached to a coiled spring hung on the inside at the top of the door frame. This was so that when the door was opened and struck the bell the owner knew he had a customer.

Pic.14. Set of horse bells with their original leather strap, late 19th-early 20th century.

Pic.15. Bell for summoning servants. It is one of a set each of which would have had a different number on it. A system of wires and bell cranks ran under the floorboards in the large homes of the wealthy, connecting the bells to individual rooms. These bells were attached to a coiled spring joined to a vertical bar, thus allowing the bell to ring. This system was still in use well into the 20th century.

Pic.16. Horse bell small example, oblong loop, 19th-early 20th century.

Open-Type Bells

Pic.17. Exterior door bells such as these are in use today. They are found mainly in rural areas, though some town houses have them as ornaments. (not to scale)

Military Items from the Napoleonic Period

When the threat of war looms, the country concerned quite naturally goes on the alert. This is not only to protect the homeland but also, if strong enough, to invade and hopefully defeat the oppressor. Such was the case when France declared war on Britain in the late 18th century, and there seemed a strong possibility of invasion.

As a result, although the French Revolutionary Wars did not begin until 1793-1802, followed by the Napoleonic Wars of 1803-1815, this country was already geared up and ready to fight.

Napoleon Bonaparte (*Napoleone Buonaparte*) was actually born in Ajaccio, Corsica, but educated in France where he attended various colleges. He eventually graduated from the military academy in 1785 and, as they say, the rest is history. As a result of the threat from France, many military camps sprung up around the country. This was especially the case in East Anglia, although many other counties established military training camps during this period (as was also the case in Ireland).

Such camps, being under canvas rather than permanent structures, could be moved very quickly. Some

Finds From a British Napoleonic Military Site

Fig.1. Officer's gilded copper button c.1795.

Fig.2. Solid cast bronze badge, single loop on back.

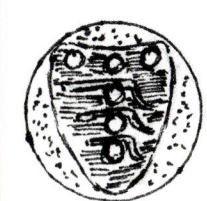

Fig.3. Copper Royal Artillery button, post 1795. There are variations known and some have been recovered with remains of gilding.

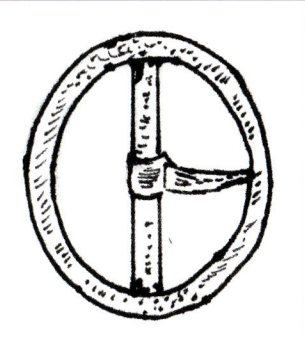

Fig.4. Copper circular buckle with pin; the centre bar protrudes from the rear.

Fig.5. Copper pressed badge with motto.

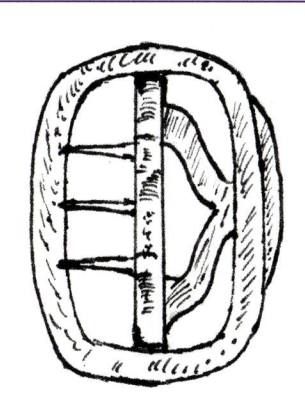

Fig.6. Stock or knap copper buckle.

Military Items from the Napoleonic Period

Fig.7. Musket flint.

Fig.8. Ornate copper shoe buckle; the inner section would have been made from iron.

Fig.9. Pistol flint.

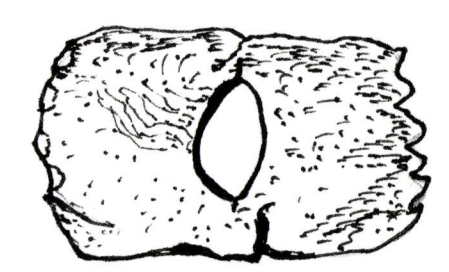

Fig.11. Lead flint holder with serrated edge.

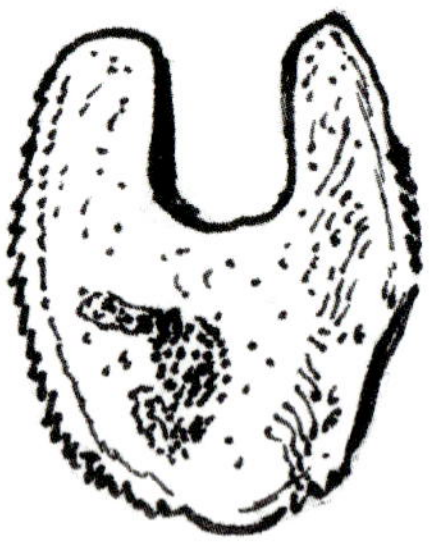

Fig.10. Lead holder for spare flint.

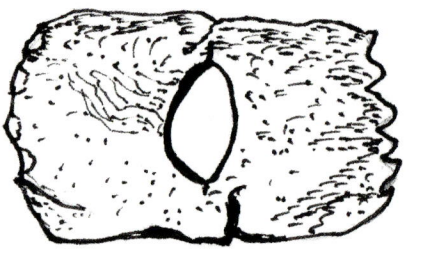

Fig.12. Lead holder for spare flint. This example is similar to Fig.10. but has a serrated edge.

Figs.13a & b. Gilded copper button with letters "LCV" (Loyal Chelmsford Volunteers).

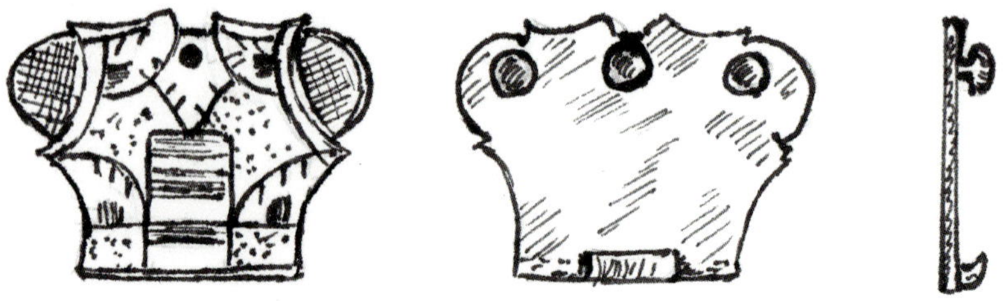

Fig.14a-c. Very ornate clasp with three studs on the back.

32

Military Items from the Napoleonic Period

temporary camps even had gun emplacements – these being sited and placed in the direction of where the enemy might come from.

Recruits drafted into these training camps for homeland defence often came from many miles away and from different parts of the UK.

On the various military sites I have searched in my own local area (Essex) I have recovered buttons from Scottish regiments that were stationed in my county, and the same situation must be mirrored around Britain.

Although originally intended to be temporary, not all such camps were simply under canvas. Some had purpose-made buildings constructed to house the officers, and a number covered over 100 acres. In fact, some training camps became virtually small villages where coffee shops supplied newspapers, and the locals were encouraged to bring along produce and other goods to sell to the soldiers.

Where camps were simply tented affairs, the officers would have found accommodations in nearby towns and villages. Some of the soldiers and officers settled and married local girls, and their descendants still remain in some villages and towns.

With so many officers and soldiers encamped in a confined area over a long period of time, a great many losses would have occurred. As might be expected, buttons are the most common items to be recovered. However, the buttons used on the uniforms of rank and file were often made of poor quality pewter. This alloy is not a good survivor in the ground as it tends to decay and flake, and if the loop was made from iron this is the first part to rot away. It is therefore rare to find such buttons in good condition, and those recovered are eagerly sought after by collectors.

Brass buttons were also widely used on the uniforms of the lower ranks, some even gilded.

Besides buttons used on the uniforms of regular soldiers, they were also used for the Volunteer and Militia groups. Such buttons were generally made from copper, with some recovered examples showing signs of gilding or silvering. It is sometimes hard to trace the regiment or group from which they derived. This is especially the case with the volunteers whose buttons usually used initial capital letters such as "LCV".

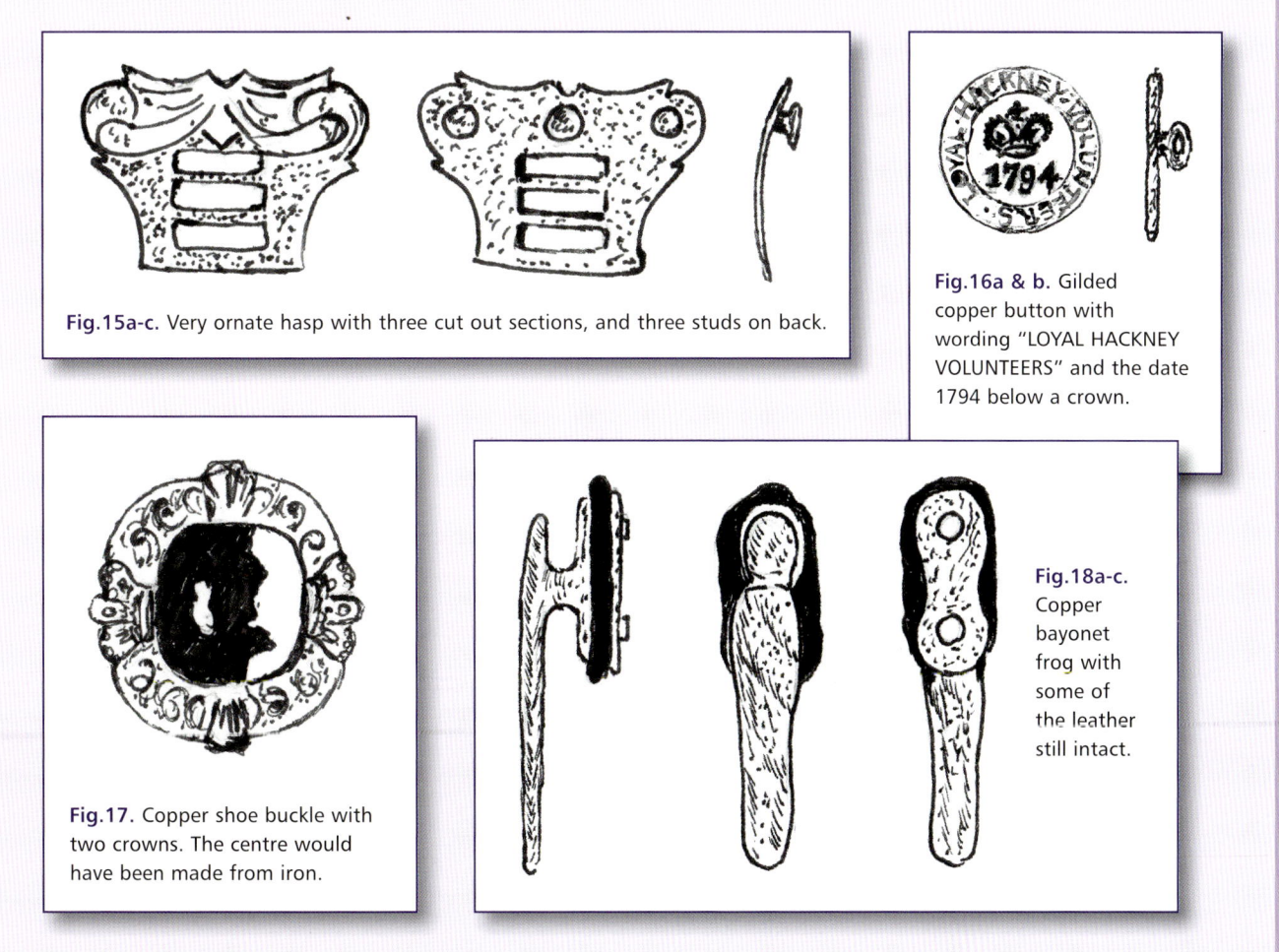

Fig.15a-c. Very ornate hasp with three cut out sections, and three studs on back.

Fig.16a & b. Gilded copper button with wording "LOYAL HACKNEY VOLUNTEERS" and the date 1794 below a crown.

Fig.17. Copper shoe buckle with two crowns. The centre would have been made from iron.

Fig.18a-c. Copper bayonet frog with some of the leather still intact.

Military Items from the Napoleonic Period

Both militia and volunteer buttons are known with a crown above the letters but this was not always used. There are various publications covering uniform buttons from the Napoleonic period, but unrecorded types are still being found.

Occasionally naval buttons are found on the sites of training camps, although – as might be expected – in much smaller numbers than army examples. Most of the examples I have found have been gilded copper (gold being resistant to salt water, sea spray etc.) The reason navy personnel might have been present could be as advisors regarding gun emplacements etc., or even recruiting.

In 1751 all regiments of the British Army were given numbers, and in 1767 the War Office ordered that each regiment of foot or dragoons should have their own particular number on their uniform buttons.

Fig.19a & b. Silver coated copper officer's button showing royal cipher.

Fig.20a & b. Dome-shaped gilded copper button of the 13th Light Dragoons.

Fig.21. Copper button of the 92nd Regiment of Foot.

Fig.22a-c. Copper fastener with suspension loop.

Fig.23. Half of snake design belt buckle.

Fig.24a & b. Copper button with letters "CM" (possibly Colchester or Chelmsford Militia).

Military Items from the Napoleonic Period

The Dragoon Guards felt that they should be exempt from this regulation, and later in the same year a new order was issued allowing them to wear buttons carrying the initials of the regiment.

Normally the buttons worn by the cavalry depended on the lace that adorned the uniform. Should the lace be silver then silver buttons would have been worn; if the lace was gold then gilded buttons would have been used.

As to the other ranks, their buttons would correspond to the colour that was worn by the officers (e.g. for yellow or white lace, brass and white metal buttons).

The buttons worn by the officers would normally have had silvered or gilded faces. The backing to which they were attached would have been made from wood or bone. Other ranks buttons were similar, though during the last quarter of the 18th century all buttons were made from metal. Buttons were either stamped or incised (engraved) with an approved number or design.

The first pattern of a Royal Artillery button dates from about 1767. This shows a cannon on its carriage facing to the right and with a pile of cannon balls in front; the whole of this design was surrounded by a rope. These buttons have a metal face but a wooden or bone backing.

In about 1785 this design was change to the shield of the Board of Ordnance. The lower ranks had bronze buttons, while the officers were gilded.

Around 1802 a new patten was worn, consisting of a crowned garter inscribed the "Royal Regt of Artillery" with inside the reversed and intertwined cipher.

In 1808 the Horse Artillery, which had been formed 1793 and had used the same design as the Royal Artillery (though on a ball button), now adopted the design of a crowned garter. This was inscribed "Royal Horse Artillery", "Royal Horse Arty" or "Royal Horse Artil". Inside this was the royal cipher. This design continued until 1855, although it was still used on the greatcoat until 1912.

The Royal Artillery changed to a different design on their button in 1833, this being a crown surmounting three guns. In 1838 this had an additional scroll added beneath, with "Ubique" and a scalloped edge. In 1855 both the Royal Artillery and Royal Horse Artillery ceased using this design and both used the crown over three guns. Both adopted a new design in 1873 showing the devise of the crown over a single field gun.

In 1803 it became apparent that the short period of peace following the Treaty of Amiens was doomed to failure, and the British Government started to prepare for another possible future conflict against Napoleon. This included the reformation of the corps of Volunteers.

During 1807 the Secretary at War, Viscount Castlereagh, considered a national reserve that would be both economical and efficient, and at the end of this year he put forward his plans for the Army and Militia. The regular Militia would consist of 120,000 men, 40,000 for Ireland and 80,000 for Britain. The Local Militia would consist of 200,000 man raised purely for local defence.

Finds From a French Napoleonic Military Site

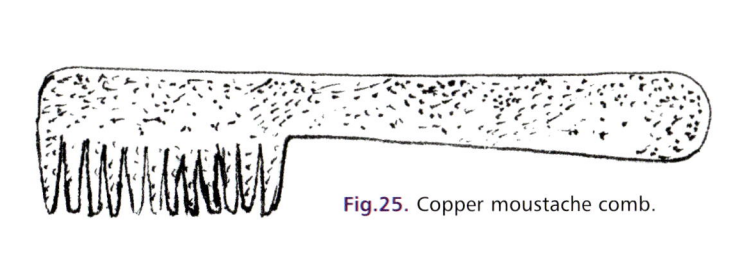

Fig.25. Copper moustache comb.

Fig.26. Iron utility knife with maker's mark on the blade. The bone handle is decorated with a copper flower.

Military Items from the Napoleonic Period

The Local Militia could be moved to any part of Britain, although this period could not exceed six weeks after the removal of the enemy. Also, they could be used in times of trouble such as to suppress riots, should these be within their county or any other adjoining county. Exercise and training was normally carried out locally, and was limited to 28 days a year.

The Local Militia was originally intended to supersede the Volunteers, and this it mostly did. However, in the City of London, the City of Westminster, and the county of Middlesex (which at that time covered parts of what is now London) where the local Volunteers were strong, the militia was never formed. It must be remembered that the area on the outskirts of the City of London was not as it is today

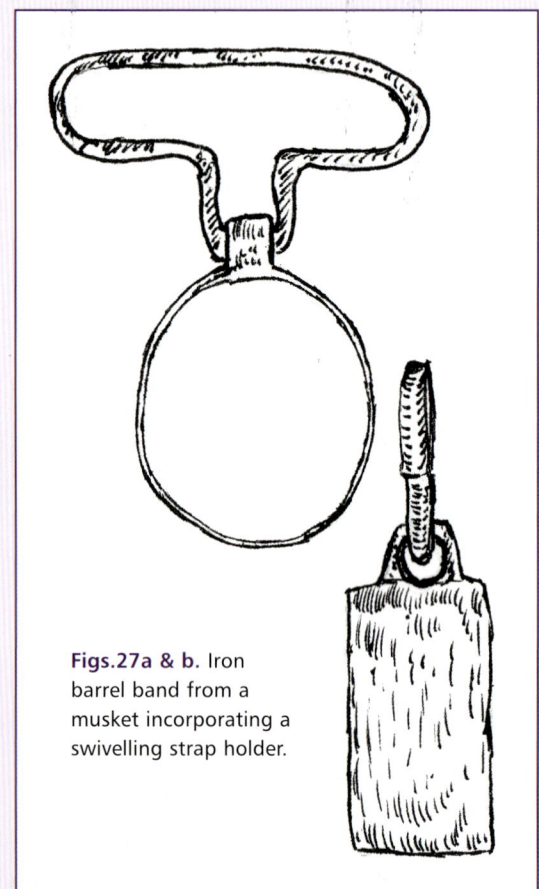

Figs.27a & b. Iron barrel band from a musket incorporating a swivelling strap holder.

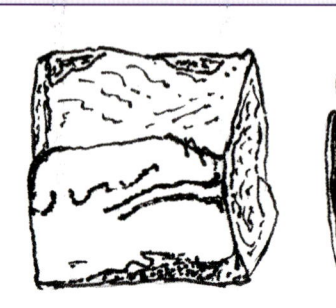 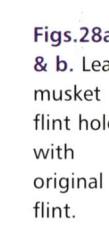

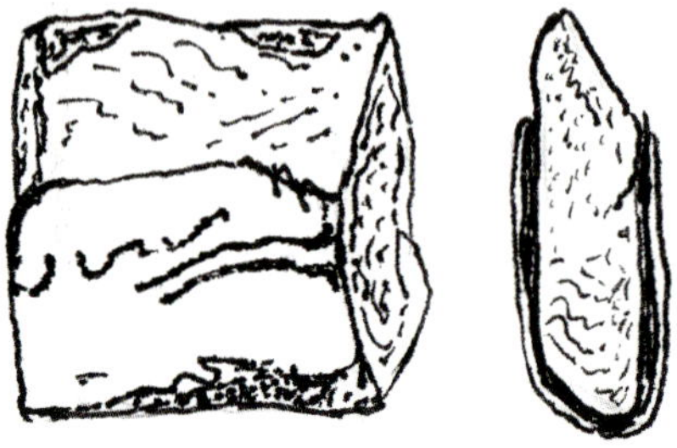

Figs.28a & b. Lead musket flint holder with original flint.

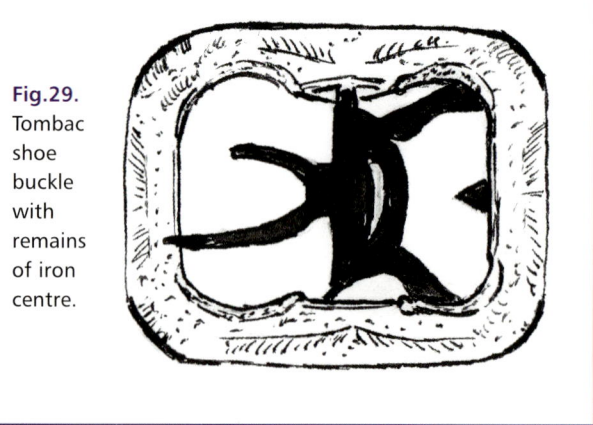

Fig.29. Tombac shoe buckle with remains of iron centre.

Figs.30a & b. French military copper button with "8" (8th Regiment) in centre surrounded by a motto. The loop is pyramid shaped, unlike those on English military buttons of the period which are round.

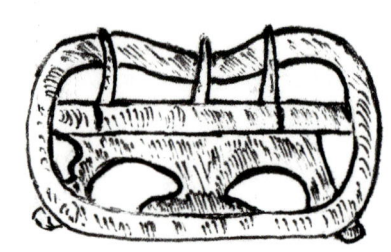

Fig.31. Complete copper buckle for stock or knap.

36

Military Items from the Napoleonic Period

and mainly consisted of villages. In the Isle of Man, Orkney, Zetland, Ireland and the Channel Islands the legislation did not apply.

The officers of the Local Militia were junior to the officers of their respective rank in the other Militia forces. The lettering on the buttons of the Volunteers normally ends in a "V". This could be in a scroll type font, as would the rest of the lettering, or in block capitals.

Some of the Volunteer buttons had the full wording around the edge within a double circle; in the centre would be the date that they were formed below a crown. One such button was of the "LOYAL HACKNEY VOLUNTEERS 1794". This company was one of those that were not replaced by the Militia. It is recorded that they acted as a reserve

Finds From British and French Napoleonic Military Sites

Pic.1. Plain copper-alloy clasp.

Pic.2. Spare flint holder with serrated edge.

Pic.3. Gilded copper naval button as worn by a captain or commander.

Pic.4. Two-pronged fork with copper ferrule on handle (French site).

Pic.5. Various stock or knap buckles (French site).

37

Military Items from the Napoleonic Period

Pic.6. Seven highly decorated machine-turned tombac buttons, all from items of clothing (French site).

Pic.7. Two copper coins (French site).

Pic.8. Musket flint (French site).

Pic.9. Tombac shoe buckle with iron inner section of single-prong type (French site).

Pic.10. Military buttons (French site).

Pic.11. Tombac buttons from French site. Note that the loops are made from iron.

Pic.12. The reverse of numbered military French buttons. On the largest example the loop is pyramid-shaped and has four holes.

Military Items from the Napoleonic Period

for which the parish watchmen could call upon for assistance when crimes were committed. They eventually fell into decline, but were re-formed in 1803 with the threat of invasion by the French; unlike the Militia, the Volunteers would normally stay in their own area and rarely venture outside of it.

There are some amusing stories regarding some of these units, and one in particular concerns the small market town of Coggeshall in the county of Essex. This place was well known for its somewhat unusual inhabitants and the peculiar things they got up to during its early period. The locals decided to band together and start a Volunteer force. This was most commendable, and all seemed well; however, the officers outnumbered the lower ranks. In fact, there was only one private. The poor fellow was illiterate and did not know his left foot from his right, but one of the officers came up with a bright idea. They tied hay to one leg and straw to the other, and marched him along calling "hay, straw, hay, straw". This seemed to work fine until others volunteered to join the ranks as privates.

The Royal Corps of Transport has an interesting and varied life. It started as the Corps of Royal Waggoners in 1794, but was disbanded in 1795. Then,

Pic.13. Silver coated copper button of the 14th Light Dragoons.

Pic.14. Various items recovered from military camp site.

Pic.15. French tombac buttons.

Pic.16. Lead spare flint holder with serrated edge (French site).

Pic.17. Musket balls. Note the holes made by using an iron worm to extract the shot after a misfire (French site).

39

Military Items from the Napoleonic Period

in 1799, they were re-raised as the Royal Waggon Corps. The name was changed once again to the Royal Waggon Train in 1802, then in 1833 they were disbanded. History soon repeated itself. Just prior to the Crimean War in 1854 they were resurrected, now called the Land Transport Corps; this changed again in 1857 to the Military Train. This title lasted until 1869 when it was altered to the Army Service Corps. As a result of their war work they were granted the prefix "Royal" in 1918, and in 1965 the RASC became the Royal Corps of Transport.

Not a great deal of coinage is recovered from the sites of former training camps, as food, clothing and accommodation was obviously provided for the lower ranks. However, money must have been needed at certain times. I know of one such camp that was situated a few hundred yards from a pub and it is documented that the establishment was frequented by soldiers; therefore they must have had some money to spend.

Badges are very rare finds from camps of this period. In Napoleonic times they were quite large, and such a loss would have soon been noticed by the wearer and the badge recovered.

Small round buckles are more frequent finds. They are similar to recoveries from the medieval period, but can be distinguished from the fact

Pic.18. Plain tombac buttons from the French site. Note that the loops differ from the British examples which are much smaller.

Pic.19. Officer's gilded copper button c.1795.

Pic.20. Solid cast bronze military badge, single loop on reverse.

Pic.21. Single-piece copper button of the Royal Artillery, post 1795. There are variations known of this example.

Pic.22. Copper circular military buckle with pin; the centre bar protrudes from the rear.

Pic.23. Pressed copper cap badge with motto.

Pic.24. Copper stock or knap buckle.

Military Items from the Napoleonic Period

that the centre bar protrudes out from the back (on the early types the centre bar is on the same plane as the rim).

Another type of buckle from military sites is the nape or stock type. On lower ranks this would have been used to hold the stiff leather collar around the neck, but with officers the collar would have been made from cotton or some more comfortable material.

Breech or knee buckles would also have been worn by some of the military. These are normally small in size ranging from about half an inch to 2 inches. As with other military items the metal from which they were made depended on whether they were worn by a rank and file soldier or an officer.

In the early 19th century shoe buckles still remained in fashionable use with the military, especially with Scottish soldiers. Typical examples are large and are normally of good quality pewter or copper with a silver coating. Such buckles would have had iron pins and inner sections, which often rust away in the ground just leaving the outer section. However, in the right ground conditions such buckles are occasionally recovered intact and some copper-alloy examples had inner sections made from the same material rather than iron.

Lead musket balls (used and unused) are common finds from military training camps, as are pistol balls

Pic.25. Musket flint.

Pic.26. Ornate copper shoe buckle; the inner section would have been made from iron.

Pic.27. Pistol flint.

Pic.29. Lead flint holder with serrated ends.

Pic.28. Lead holder for spare musket flint.

Pic.30. Very ornate clasp, reverse showing three studs.

Pic.31. Gilded copper button of the "Loyal Hackney Volunteers" with date 1794 below a crown.

41

Military Items from the Napoleonic Period

(from the weapons owned and used by officers). Musket and pistol flints can sometimes be found "eyes only" on ploughed farmland sites, or sometimes with the aid of detectors when they were lost still accompanied by their lead protective holders.

Musket parts can occasionally be found, but these are rare for the individual soldier was responsible for the care and maintenance of his gun and would face serious charges if pieces went missing.

A group of the items illustrated were actually recovered from a Napoleonic military camp on the Continent. From these it would seem that the French experienced similar types of artefact losses to the British. However, one item of particular interest is the moustache comb. These are extremely rare items to be recovered from British camps, but turn up fairly frequently on the sites of French camps. One French custom concerning the Napoleon's Old Guard was that between 1 March and 1 December each year the moustache was sported; however, during the winter it was shaved off. After 1807 this custom was changed and the moustache was worn throughout the year.

Searches of these Napoleonic military camp sites also produce numerous civilian items. Many outsiders would have visited the camps – whether for social or business reasons – and they would also have lost various items over a period of time.

Pic.32. Copper shoe buckle with crown at each end; the central would have been made from iron.

Pic.33. Copper bayonet frog with some of the original leather still intact.

Pic.34. Officer's silvered copper button showing royal cipher.

Pic.35. One half of a belt buckle, the fastener being of snake design.

Pic.36. Copper fastener with suspension loop.

42

Buckle Plates

Buckle plates have been in use for many centuries. They range from those of plain design to examples that are almost miniature works of art. In the type of metals from which they were made, this can vary from humble copper-alloy through to gold; this also applies to the buckles to which they were (or still are) attached.

The buckles in many cases – and even those of precious metal – would have been plain as opposed to the ornate attached plate. In cases where both were decorated, however, they are a sight to behold.

Unfortunately, when lost and suffering burial in the ground, the plates tend to part company from the buckles. Some plates were made quite thin, and normally only secured to the buckle by two small shoulders that also, in some types, acted as hinges.

However, there are some early

Fig.1. Lion's head buckle and plate. The plate has two raised central lines and four rivets, late Roman.

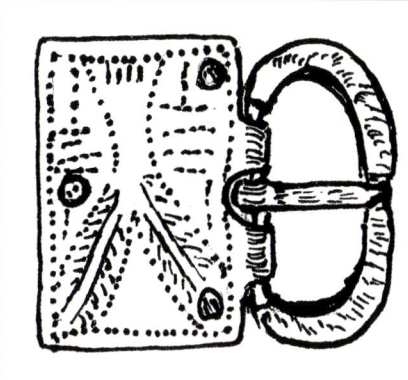

Fig.2. Bronze buckle and plate decorated in a dot design in the form of leaves, 4th century.

Fig.3. Bronze buckle and plate, dot design in the form of a bird within vertical lines, single rivet, 4th century.

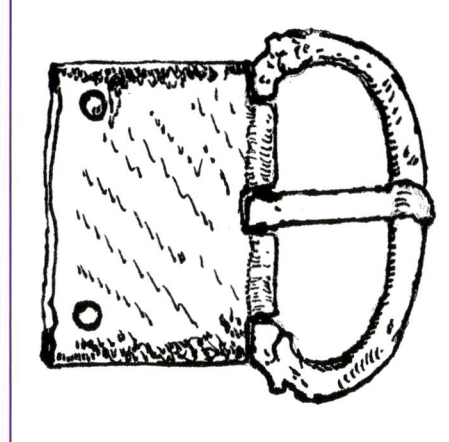

Fig.5. Bronze buckle (dragon type) and plate, plain rivets, late 4th century.

Fig.4. Bronze buckle, incomplete, from body armour.

Fig.6. Bronze plate, hinged type from body armour, three rivets.

Buckle Plates

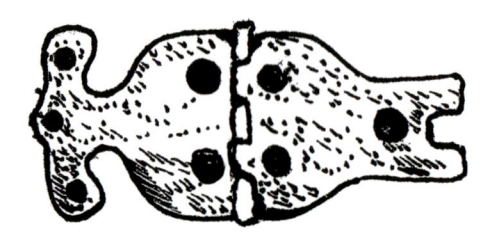

Fig.7. Bronze hinged plate, buckle missing, from body armour.

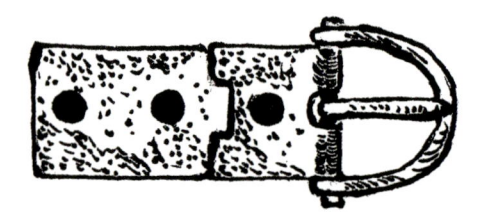

Fig.8. Bronze complete hinged buckle and plate, from body armour, three rivets.

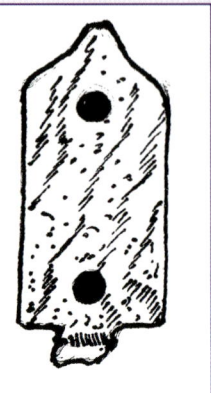

Fig.9. Bronze section of hinged plate from armour.

Fig.10. Bronze crested buckle with fixed plate, openwork design, 4th century.

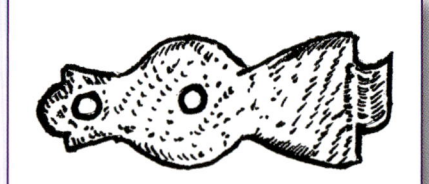

Fig.11. Bronze section of hinged plate from body armour.

Fig.12. Bronze plate with cross-hatched design, 4th century.

Fig.13. Bronze openwork buckle plate, section missing, late Roman.

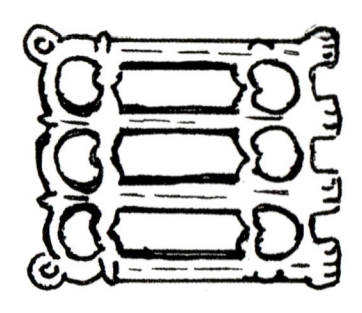

Fig.15. Bronze openwork buckle, plate intact, late Roman.

Fig.14. Bronze openwork buckle, section missing, late Roman.

Buckle Plates

examples (dating from the Roman, Saxon or early medieval periods) where the buckle plates were made of much thicker metal or even moulded as an integral part. Such examples tend to remain intact and have a much better chance of survival in the ground.

During the Roman Occupation of Britain various styles of buckles and plates were in use, some originating from local craftsmen or manufacturing workshops. Continentally-derived buckles and plates have also been recovered in this country, no doubt originating from foreign recruits or mercenaries when the Romans needed a larger force in troubled times, or in the early 5th century when the Roman army was withdrawn.

For a more in-depth look at Roman buckles and plates I can highly recommend **Roman Buckles & Military Fittings** by Andrew Appels and Stuart Laycock (available from Greenlight Publishing).

Items from any part of the Saxon period, as one might expect, are rare.

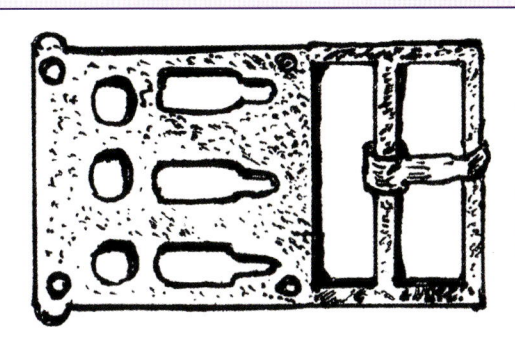

Fig.16. Bronze fixed openwork buckle and plate, late Roman.

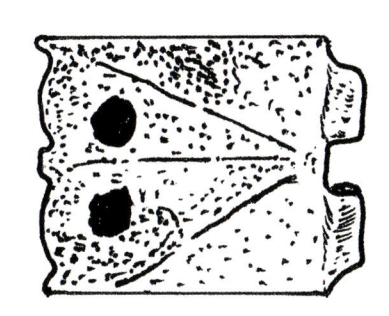

Fig.17. Bronze rectangular plate, three radiate lines decoration and scalloped edge, 4th century.

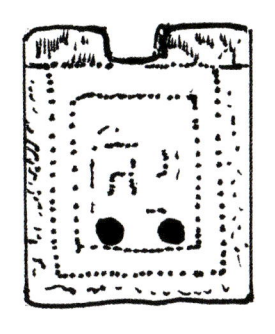

Fig.18. Bronze rectangular plate with repoussé swastika design, 4th century.

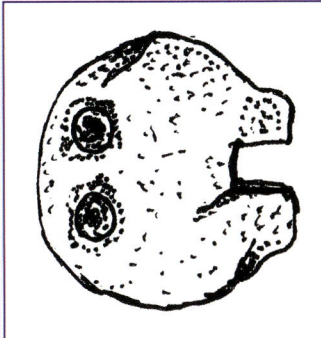

Fig.19. Bronze circular plate, two rivets, 4th century.

Fig.20. Bronze rectangular plate, scalloped edges, two rivets, 4th century.

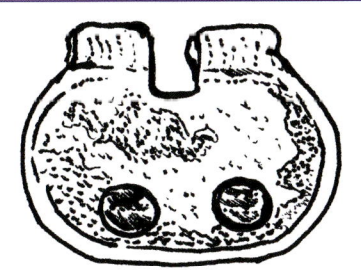

Fig.21. Bronze circular plate, outer lined edge, two rivets, 4th century.

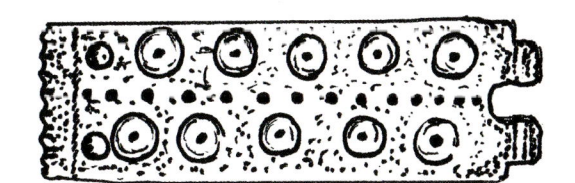

Fig.22. Bronze oblong plate, ring and dot centre with dotted lines, scalloped edge, 4th century.

Buckle Plates

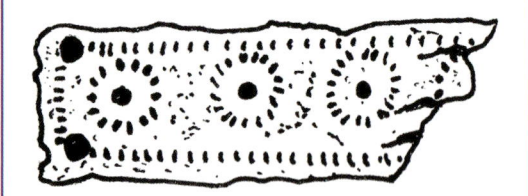

Fig.23. Bronze section of oblong plate decorated with repoussé rosettes and lined border, 4th century.

Fig.24. Bronze rectangular plate, single large ring and dot, two rivets, 4th century.

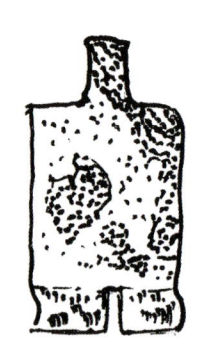

Fig.25. Bronze section of hinge from body armour.

Fig.26. Bronze hinge plate, tri-lobed type, from body armour.

Fig.29. Saxon bronze buckle and plate, cast in one piece, central design and single domed-shaped rivet, 8th century.

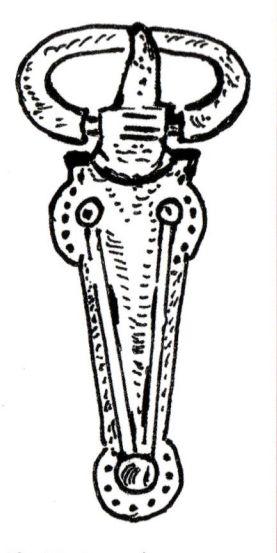

Fig.27. Saxon bronze buckle and plate, the triangular plate with transverse lines, single domed-shaped rivet, 5th-6th century.

Fig.28. Saxon bronze buckle and plate, triangular plate with transverse lines and ring and dot, three domed-shaped rivets, 6th century.

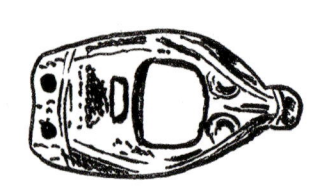

Fig.30 Saxon bronze buckle and plate cast in one piece, transverse lines, two small rivets, 9th century.

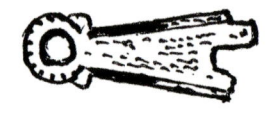

Fig.31. Saxon bronze plate, transverse lines, single rivet (missing), 6th century.

Buckle Plates

The population during the latter part of the Saxon England period was approximately 1 million. This might seem a great deal of people, but if they had all been gathered together they would only be about the same amount that live in modern day Birmingham. Also, many people lived in small communities possibly numbering only 10 to 20 individuals. Farming was their main source of income, although trading in wool was also known. As a result, Saxon items are much sought after – not only be collectors but also historians.

Some Saxon buckle plates, as with other styles from various periods, can have a folded plate that was added to the buckle after manufacture. The alternative was an integral (rigid) plate with both buckle and plate cast as one piece. The integral type appears to have been in particular favour during the Saxon period as a number of examples

Fig.32. Saxon bronze buckle and plate, single piece, two rivets (missing), 9th-10th century.

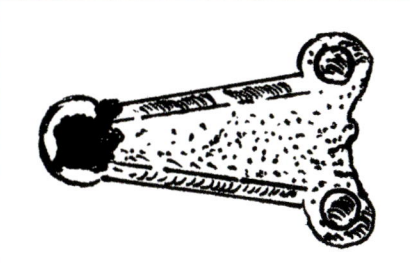

Fig.33. Saxon bronze triangular plate, transverse lines, three rivets remaining (two domed-shaped made of bronze, one rusted made of iron), 6th century.

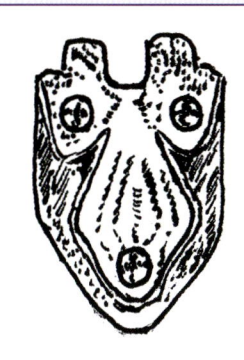

Fig.34. Saxon bronze plate, transverse lines in the shape of a fleur-de-lis, three dome-shaped rivets, possibly 10th century.

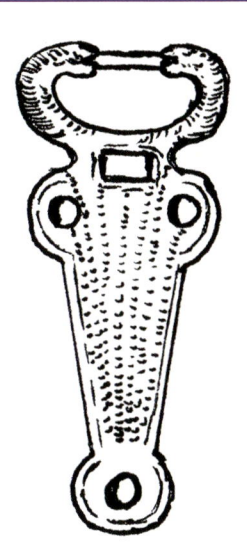

Fig.35. Saxon single-piece bronze buckle and plate, triangular plate, transverse lines, three rivet holes, 6th century.

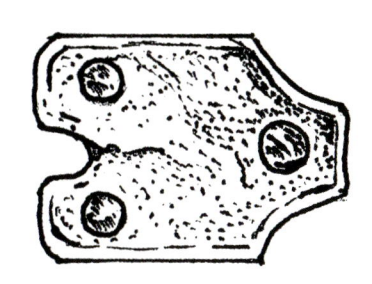

Fig.36. Saxon bronze plate, plain with three domed-shaped rivets, possibly 10th century.

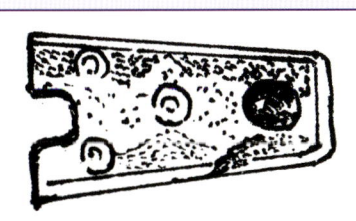

Fig.37. Saxon bronze plate with three ring and dots, single domed-shaped rivet, possibly 9th-10th century.

Fig.38. Saxon single-piece bronze buckle and plate, triangular shaped, ring and dot design, single dome-shaped rivet, 6th century.

Buckle Plates

Fig.39. Saxon single-piece buckle and plate, central design single rivet missing, 8th century.

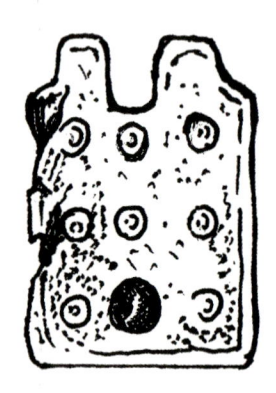

Fig.40. Saxon bronze plate, ring and dot design, single dome-shaped rivet, possibly 9th-10th century.

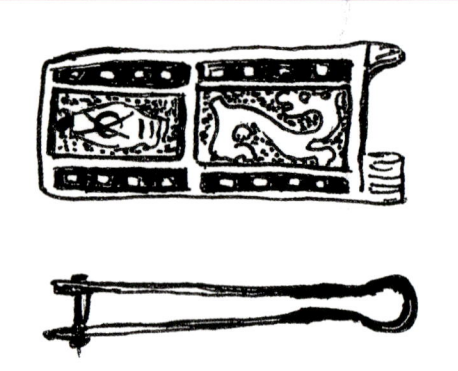

Fig.41. Saxon bronze plate "Trewhiddle style" (so named after a hoard from Cornwall).

Fig.42. Norman bronze plate, mythical beast, 11th century.

Fig.43. Norman bronze plate, mythical beast 11th century.

Fig.44. Viking/Saxon buckle and plate cast in one piece, abstract design, 9th-10th century.

Fig.45. Viking/Saxon plate, unusual abstract design 9th-10th century.

Fig.46. Viking bronze plate, floral design, 9th-10th century.

Fig.47. Viking bronze buckle and plate cast in one piece, interlaced design, 9th-10th century.

Buckle Plates

have been unearthed both by detector users and during archaeological excavations. In the main these were made from copper alloy, although some were cast from precious metal.

Some of the folding buckle plates carry elaborate designs in the form of mythical creatures, though when buckle and plate are recovered intact the buckle is generally plain. One of the most well-known examples of buckle and plate was recovered at the Sutton Hoo excavations just before the outbreak of the Second World War. This example was made from solid gold and has been the discussion point of many previous articles and publications. Another famous buckle and plate is the "Alton Buckle" this having been discovered during excavations of an Anglo-Saxon cemetery during the late 1950s to early 1960s. This has a silver body of sub-triangular shape with filigree wire and niello set with garnets

Fig.48. Saxon bronze plate similar to Fig.41.

Fig.49. Norman bronze buckle and plate, mythical beast, 11th century.

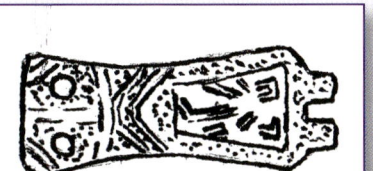

Fig.51. Viking/Saxon bronze plate, dog head type, 10th-11th century.

Fig.53. Viking bronze plate, howling beast, 10th-11th century.

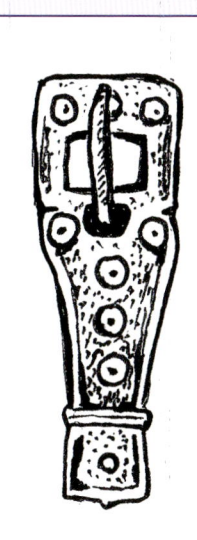

Fig.50. Saxon buckle and plate cast in one piece, ring and dot design, 7th-8th century.

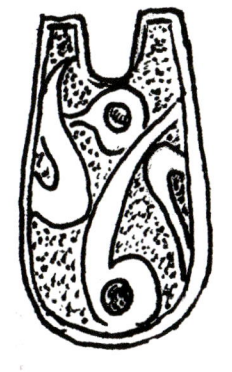

Fig.52. Viking bronze plate, floral design, 9th-10th century.

Fig.54. Viking/Saxon buckle and plate, abstract design, 10th-11th century.

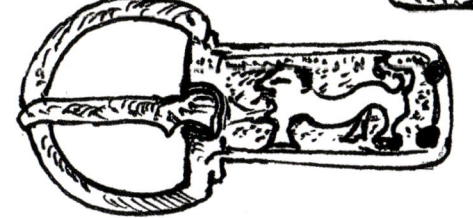

Fig.55. Norman bronze buckle and plate, lion within a lined border, 11th-12th century.

49

Buckle Plates

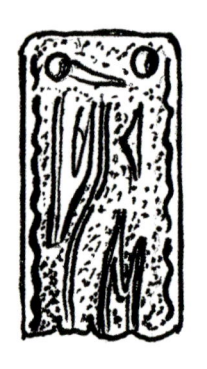

Fig.56. Viking/Saxon plate (incomplete), abstract design, 10th-11th century.

Fig.57. Saxon bronze plate, possibly some form of fish within a lined border, 10th-11th century.

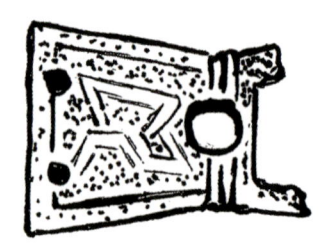

Fig.58. Medieval buckle and plate cast in one piece (section of buckle missing), integral box chape, late 14th-early 15th century.

Fig.61. Norman bronze plate, mythical creature facing right, 11th-12th century.

Fig.60. Norman bronze plate, openwork design depicting a facing lion, 11th-12th century.

Fig.59. Norman bronze plate, line design, 11th-12th century.

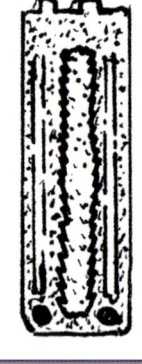

Fig.62. Norman bronze plate, possibly a lion (somewhat worn), 11th-12th century.

Fig.63. Medieval plate, diamond pattern with four outer leaves mirrored in the lower half, 15th century.

Fig.64. Medieval fastener plate, four circles each having five dots, two crosses on rear, 15th century.

Fig.65. Norman bronze buckle depicting coronation scene, Romanesque type; there are similar examples showing three people, 12th century.

Buckle Plates

and glass. The centre panel of semi-zoomorphic design is gold filigree on a repoussé base.

Such buckles, regardless of the metal from which they were made, were intended to be seen and admired. As with many artefacts, the adoption of style and design from one culture to another was an often lengthy process and not achieved over a short period of time.

This also applies to the Norman Invasion of England in 1066. Although William of Normandy defeated the Saxons, and thereby introduced a new way of life and dress, the style or fashion did not suddenly cease. In fact, some of the existing fashions carried through, on both sides, right up until the 12th century.

The most widely-used buckle favoured by the Normans and also used by the Saxons was the "D" shaped type. This remained in use

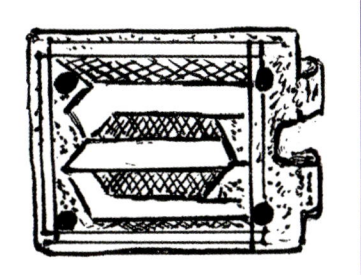

Fig.66. Medieval plate, black letter "M" on cross-hatched background, 14th-early 15th century.

Fig.67. Medieval plate showing three lions, 13th-14th century.

Fig.68. Norman bronze plate, figure "8" shaped swirl design, 11th-12th century.

Fig.69. Buckle plate showing an engraved mythical creature surrounded by a lined border, two rivets, 1350-1450.

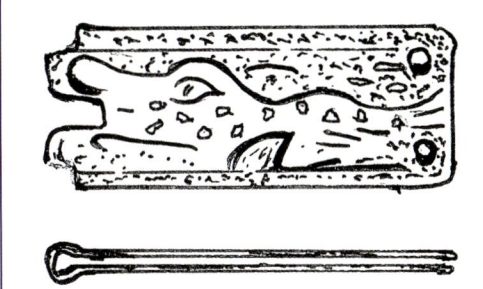

Fig.70. Plate showing an engraved mythical creature surrounded by a lined border, two rivet holes, 1350-1450.

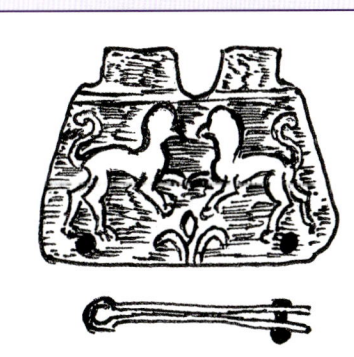

Fig.72. Plate showing two creatures facing each other with a fleur-de-lis below, two rivet holes, 1250-1350.

Fig.71. Part of a plate showing a mythical creature, 1350-1450.

Buckle Plates

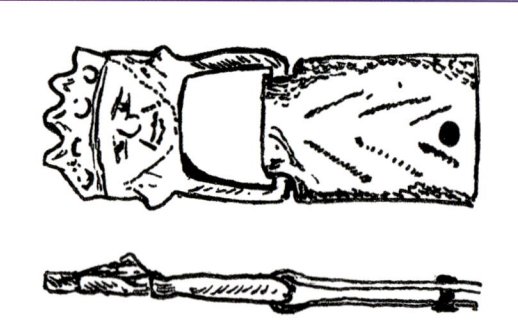

Fig.73. Frame and plate, the frame showing a crowned face while the plate is plain, one rivet hole, 1300-1400.

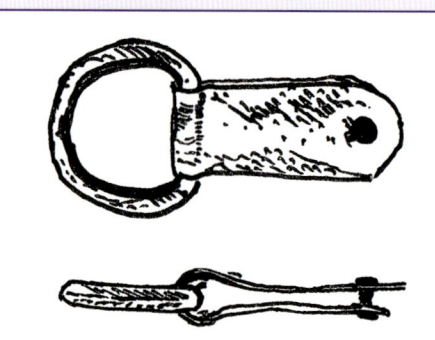

Fig.74. Oval frame and plain plate, one rivet hole, 1300-1400.

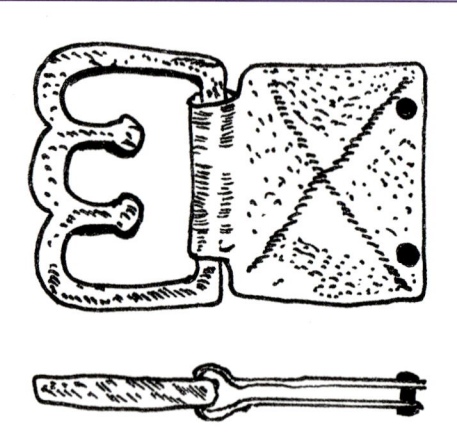

Fig.75. Frame and plate, the frame having two internal projections while the plate has an "X" done in wriggle work, two rivet holes, 1400-1500.

Fig.76. Frame plate with wriggle work lines, 1400-1500.

Fig.79. Plate showing leopard looking backwards surrounded by a double lined border, 1250-1400.

Fig.78. Plate showing a bird (possibly a dove), surrounded by a lined border, two rivet holes, 1250-1400.

Fig.77. Plate showing dog or wolf surrounded by a lined border, three rivet holes, 1250-1400.

Buckle Plates

through many centuries, and is still in use today.

The mid 13th century through to the late 14th century is the period when buckle plates seemed to have been used more than any other period in history judging from the amount that has been found by detectorists. In the main these are of the plain, thin folded sheet types having between one or two to five rivets. Those with minimal decoration normally have a border. This could consist of a straight line, punched dot, beaded or crudely engraved wriggle work design. Others had a single or double cross or transverse and radiating lines. Apart from the usual lines some would be inscribed with lettering such as "AVE LEU" and show a mythological beast. Die stamping was used on some buckles, the field being stippled by multiple punching. Raised rivets appear to have been popular with this type of plate, with both buckle and plate gilded.

However, not all plates were engraved or die stamped. Some would have been secured with small or large

Fig.80. Plate showing an elaborate geometrical design surrounded by a lined border, two rivets, 1300-1450.

Fig.81. Plate showing what is possibly a leopard surrounded by a double line border with radiate lines, two rivet holes, 1250-1400.

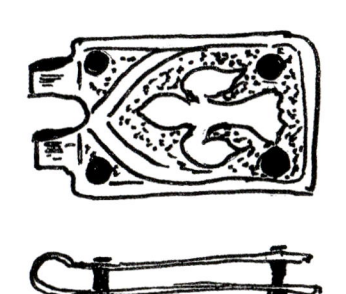

Fig.82. Plate showing a fleur-de-lis within an arch surrounded by a lined border, four rivet holes, 1250-1400.

Fig.83. Plate of unusual design showing three human figures; the large hole would indicate that this was a repair, two original rivet holes visible, 1250-1400.

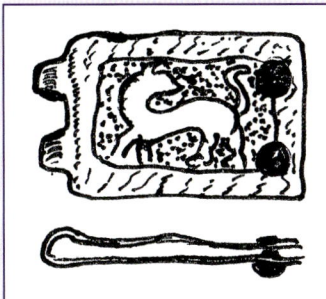

Fig.84. Plate showing leopard looking backwards, two rivet holes, 1250-1400.

Fig.85. Plate of unusual geometric design, two rivet holes, 1250-1400.

Fig.86. Plate consisting of two lines of dots, two rivet holes, 1350-1450.

Buckle Plates

Fig.87. Plate showing leopard facing, two rivet holes, 1250-1400.

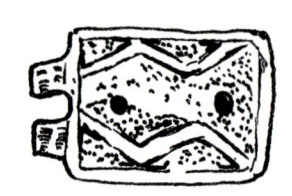

Fig.88. Plate showing double-lined design surrounded by a lined border, two rivet holes, 1300-1450.

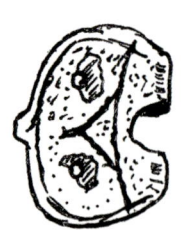

Fig.89. Plate of oval shape showing single lines, two rivet holes, 1450-1600.

Fig.90. Plate dotted border, three rivet holes, 1500-1600.

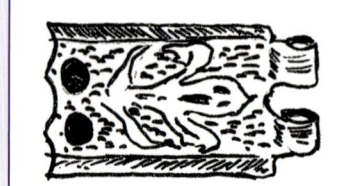

Fig.91. Plate with design showing crude fleur-de-lis, two rivet holes, 1250-1600.

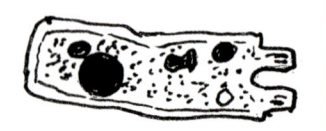

Fig.92. Plate with lined border. This example has five rivet holes two of which are crudely made (in particular the larger one) these being added at a later date, 1250-1600.

Fig.95. Plate, plain, two rivet holes, 1450-1600.

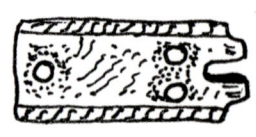

Fig.94. Plate with rope edge design, three rivet holes, 1450-1600.

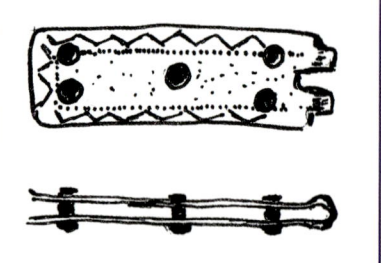

Fig.96. Plate with beaded line surrounded by a wriggle-work design, five rivet holes, 1450-1600.

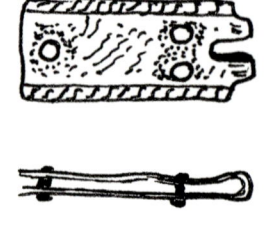

Fig.93. Plate with some of the buckle still intact, five rivet holes, 1250-1500.

Buckle Plates

decorated cast studs. These added a decorative purpose while still serving the practical use of attaching the plate to the belt.

Flat single integral plates were also popular during the medieval period. These had a varied use. Those terminating in a hook are normally associated with the spur buckles (although flat examples with rivet holes were also used). When the loop on this type of buckle is damaged or broken off it can often be mistaken for one of a different type.

Another type of integral plate had a limited life during the late 14th to early 15th centuries. These had a box plate normally engraved with some type of pattern or religious lettering such as "IHC". These were in favour for only a short period of time.

The single loop buckle plate with integral fork spacers was popular from the mid 14th to mid 15th centuries. It had two separate sheets soldered onto the fork each having one or two rivet holes.

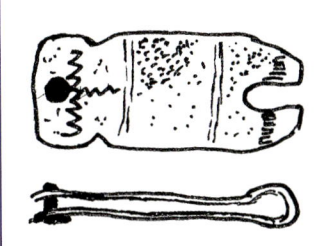

Fig.97. Plate with line and wriggle-work design, 1450-1600.

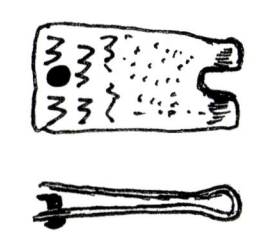

Fig.98. Plate with wriggle-work design, one rivet, 1450-1600.

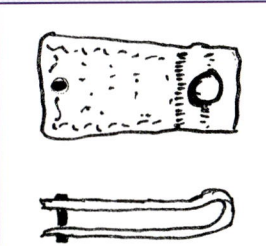

Fig.99. Plate with somewhat worn design, one rivet, 1400-1600.

Fig.100. Plate with beaded border, two rivet holes, 1450-1600.

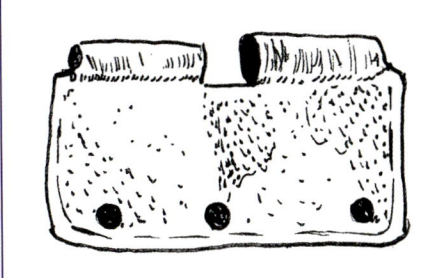

Fig.101. Large plate, lined border, three rivet holes, 1450-1600.

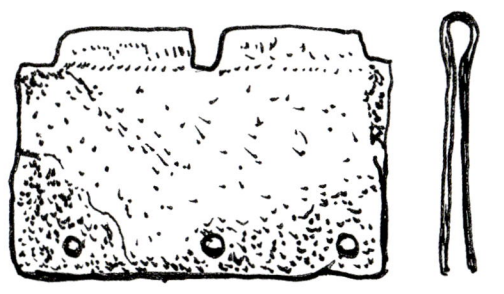

Fig.103. Large plate, plain, three rivet holes, 1450-1600.

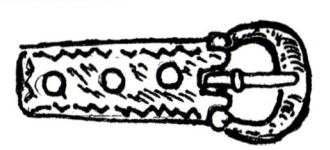

Fig.104. Buckle and plate with wriggle-work border, three rivet holes, 1250-1500.

Fig.102. Large plate, lined border surrounded by wriggle work, three rivet holes, 1450-1600.

Buckle Plates

Fig.105. Buckle and plate plain with five rivet holes, 1250-1500.

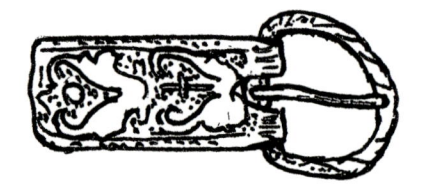

Fig.106. Buckle and plate, floral design surrounded by a lined border, one rivet hole, 1350-1500.

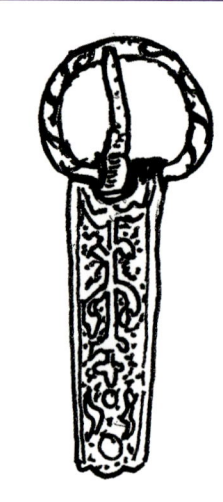

Fig.107. Buckle and plate, floral design, one rivet hole, 1350-1500.

Fig.108. Buckle and plate, floral design, one rivet hole, 1350-1500.

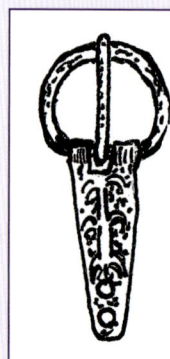

Fig.109. Buckle and plate, floral design, two rivets, 1350-1500.

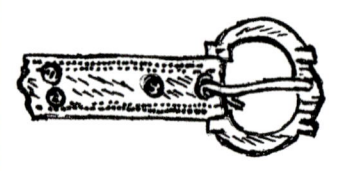

Fig.110. Buckle and plate, beaded border, three rivet holes, 1250-1500.

Fig.111. Buckle and plate inscribed "AVA LEV", two rivet holes, 1250-1400.

Fig.112. Buckle and plate, plain, three rivet holes, 1250-1400.

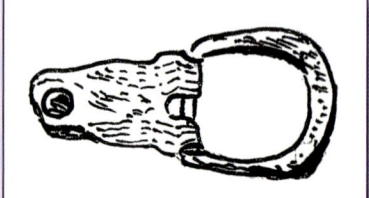

Fig.115. Buckle and plate, plain, one rivet hole, 1250-1400.

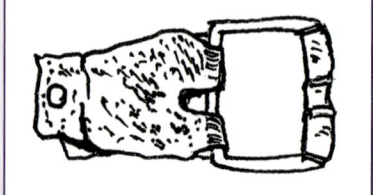

Fig.114. Buckle and plate, plain, one rivet hole, 1250-1400.

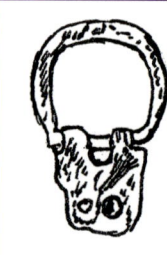

Fig.113. Buckle and plate, plain, two rivet holes, 1250-1400.

Buckle Plates

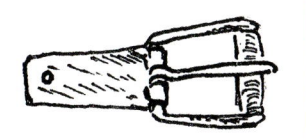

Fig.116. Buckle and plate, plain, one rivet hole, 1250-1400.

Fig.117. Buckle and plate, plain, two rivet holes, 1250-1400.

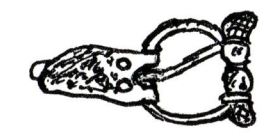

Fig.118. Buckle and plate, plain, two rivet holes, 1250-1400.

Fig.119. Buckle and plate, plain, one rivet hole, 1250-1450.

Fig.120. Buckle and plate, lined border, three rivet holes, 1250-1400.

Fig.121. Buckle and plate, floral design, two rivets, 1480-1650.

Fig.122. Buckle and plate, plain, three rivet holes, 1250-1400.

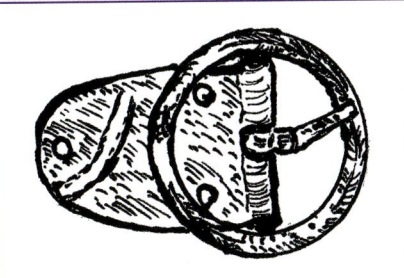

Fig.124. Buckle and plate, crescent shaped with double lines, three rivet holes, 1350-1500.

Fig.123. Large buckle and plate made from iron. The plate is plain and has one large rivet hole, 1350-1500 (scaling 60%).

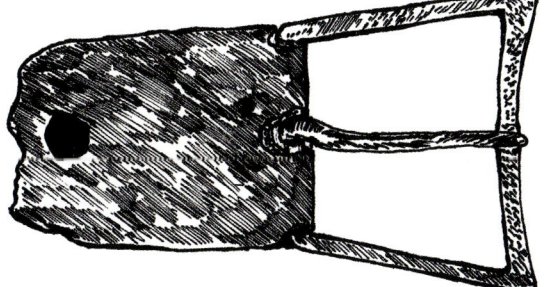

Fig.125. Buckle and plate, double lined border, three rivet holes, 1400-1500.

57

Buckle Plates

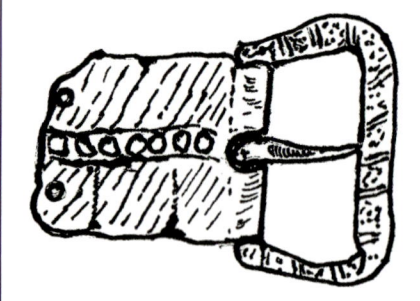

Fig.126. Buckle and plate, double horizontal lines with circles, two rivets, 1350-1500.

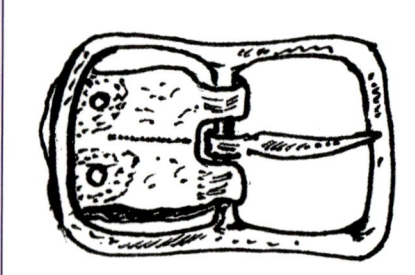

Fig.127. Buckle and plate, plain, two rivet holes, 1400-1550.

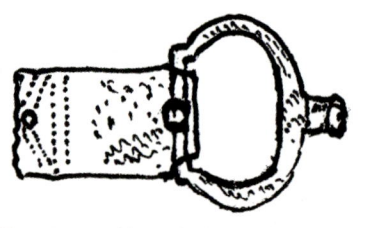

Fig.128. Buckle and plate, double dotted lines, one rivet hole, 1350-1450.

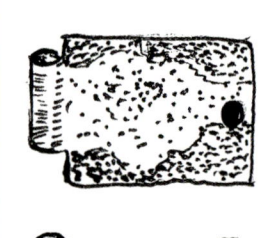

Fig.129. Buckle and plate, plain, 1400-1600.

Fig.130. Buckle/clasp and plate, faint zigzag lines, two rivet holes, 1350-1450.

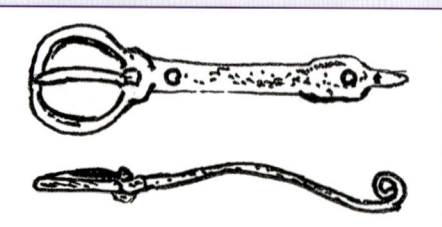

Fig.132. Buckle and plate showing a bird surrounded by a lined border, 1250-1400.

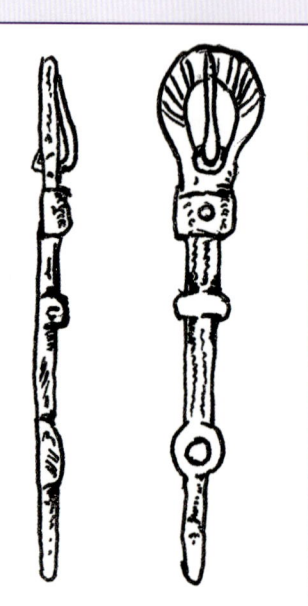

Fig.134. Medieval copper-alloy spur buckle with integral plate, transverse ridge, two rivet holes, 1250-1500.

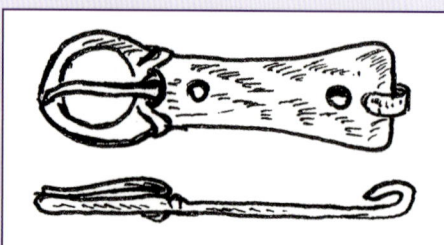

Fig.133. Medieval copper-alloy spur buckle with integral plate terminating in a loop, two rivet holes, 1250-1500.

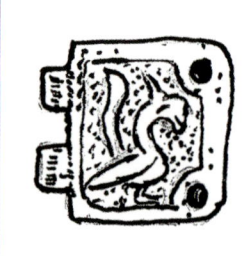

Fig.131. Buckle and plate, mythical creature (possibly a bird), two rivet holes, 1200-1350.

Buckle Plates

Fig.135. Medieval copper-alloy spur buckle with integral plate terminating in a transverse ridge with a serrated end, one rivet hole, 1250-1500.

Fig.136. Medieval copper-alloy spur buckle with integral plate, plain, one rivet hole, 1250-1500.

Fig.137. Medieval copper-alloy spur buckle with integral plate, holed centre, two rivet holes, 1250-1500.

Pic.1. Buckle and plate with wriggle-work border, three rivet holes, 1250-1300.

Pic.2. Buckle and plate, plain, three rivet holes, 1250-1400.

Pic.3. Buckle and plate, floral design, two rivet holes, 1350-1500.

Figs.138-145. Parts of medieval integral plates.

59

Buckle Plates

Pic.4. Buckle and plate, plain, one rivet hole, 1250-1400.

Pic.5. Buckle and plate, floral design, two rivet holes, 1350-1500.

Pic.6. Buckle and plate, floral design surrounded by a lined border, one rivet hole, 1350-1500.

Pic.7. Buckle and plate, beaded border, three rivet holes, 1250-1500.

Pic.8. Buckle and plate inscribed "AVA LEU", two rivet holes, 1250-1400.

Pic.9. Cast buckle with integral plate, two rivets, 1250-1500.

Pic.10. Buckle and plate, plain (still retains some of the original cloth in the plate), 1350-1450.

Pic.11. Buckle and plate, plain, one rivet hole, 1250-1400.

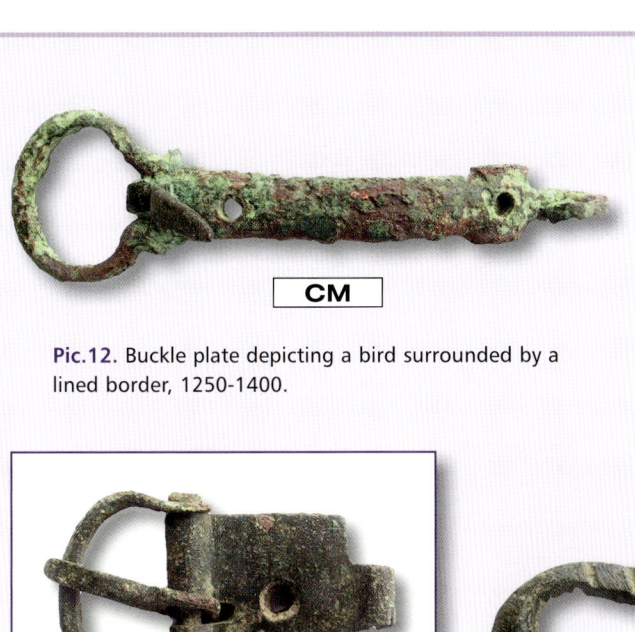

Buckle Plates

Pic.12. Buckle plate depicting a bird surrounded by a lined border, 1250-1400.

Pic.13. Buckle plate with part of the buckle remaining, five rivet holes, 1250-1500.

Pic.14. Roman bronze buckle (incomplete) from body armour.

Pic.15. Buckle and plate, double horizontal lines with row of circles, two rivet holes, 1350-1500.

Pic.16. Buckle and plate, floral design, two rivets, 1480-1650.

Pic.17. Buckle and plate, double lined crescent design, three rivet holes, 1350-1500.

Pic.18. Buckle plate plain but with remains of gilding, five rivet holes, 1450-1600.

61

Buckle Plates

Pic.19. Buckle plate with lined border. This example has five rivets two of which are crudely made (in particular the large one), these having been added after manufacture, 1250-1500.

Pic.20. Buckle and plate, plain, with much of its original gilding remaining, five rivets, 1250-1500.

Pic.21. Buckle and plate with wriggle-work border and about 95% of its gilding remaining, three rivets, 1250-1500.

Pic.22. Buckle and plate, plain, two rivet holes, 1400-1550.

Pic.23. Buckle plate showing an elaborate geometrical design surrounded by a lined border, two rivet holes, 1300-1450.

Pic.24. Buckle and plate, double lined border, three rivet holes, 1400-1500.

Buckle Plates

Pic.25. Buckle plate of oval shape design showing single lines, two rivet holes, 1450-1600.

Pic.27. Buckle plate, lined border, signs of gilding, five rivet holes, 1350-1500.

Pic.26. Buckle/clasp and plate, faint zigzag lines, two rivet holes, 1350-1450.

Pic.28. Buckle plate with unusual geometric design, two rivet holes, 1250-1500.

Pic.29. Cast buckle with small integral plate, plain, one rivet hole, 1250-1500.

Buckle Plates

Pic.30. Buckle and plate, plain, one rivet hole, 1250-1400.

Pic.31. Buckle and plate showing remains of tinning, and some cloth intact in the plate, plain, 1350-1450.

Pic.32. Buckle and plate, some tinning remaining, plain, one rivet hole, 1350-1450.

Pic.33. Buckle and plate, plain, two rivet holes, 1450-1600.

Pic.34. Buckle and plate, plain, two rivet holes, 1250-1400.

Pic.35. Buckle and plate, plain, one rivet hole, 1350-1450.

Pic.36. Buckle and plate, plain, one rivet hole, 1350-1500.

Pic.37. Late Roman lion head buckle and plate; the plate has two raised lines and four rivets.

Belt Chapes

The belt chape was (and still is) a terminal secured to the end of a belt to prevent it from fraying. It was also an item that eventually became a fashion accessory that has lasted in favour down through the centuries. By the 4th to 5th centuries the Roman military belt chape was in wide use and such items were highly prized.

The most popular style was of amphora shape, some examples being plain but in the majority of cases carrying some form of design. Some of this decoration was in the form of ring and dots that varied in size and in themselves were used to form a pattern. Other chapes had wriggle work or punched beaded design or a combination of two of these styles or all three. Some chapes were also formed with openwork – this often being of crescent shape.

Chapes varied in size although in the main they were reasonably large for,

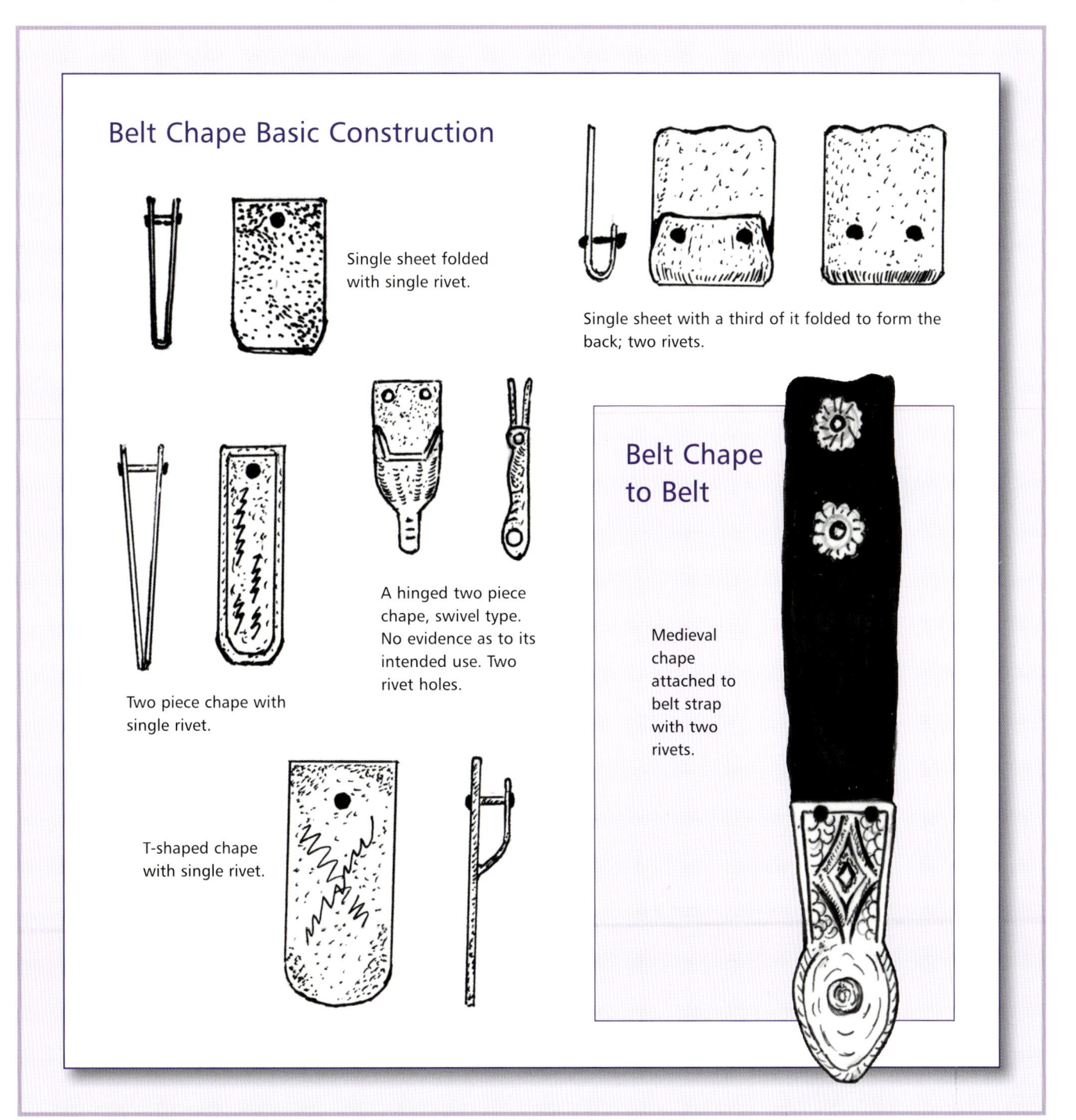

Belt Chape Basic Construction

Single sheet folded with single rivet.

Single sheet with a third of it folded to form the back; two rivets.

Two piece chape with single rivet.

A hinged two piece chape, swivel type. No evidence as to its intended use. Two rivet holes.

T-shaped chape with single rivet.

Belt Chape to Belt

Medieval chape attached to belt strap with two rivets.

65

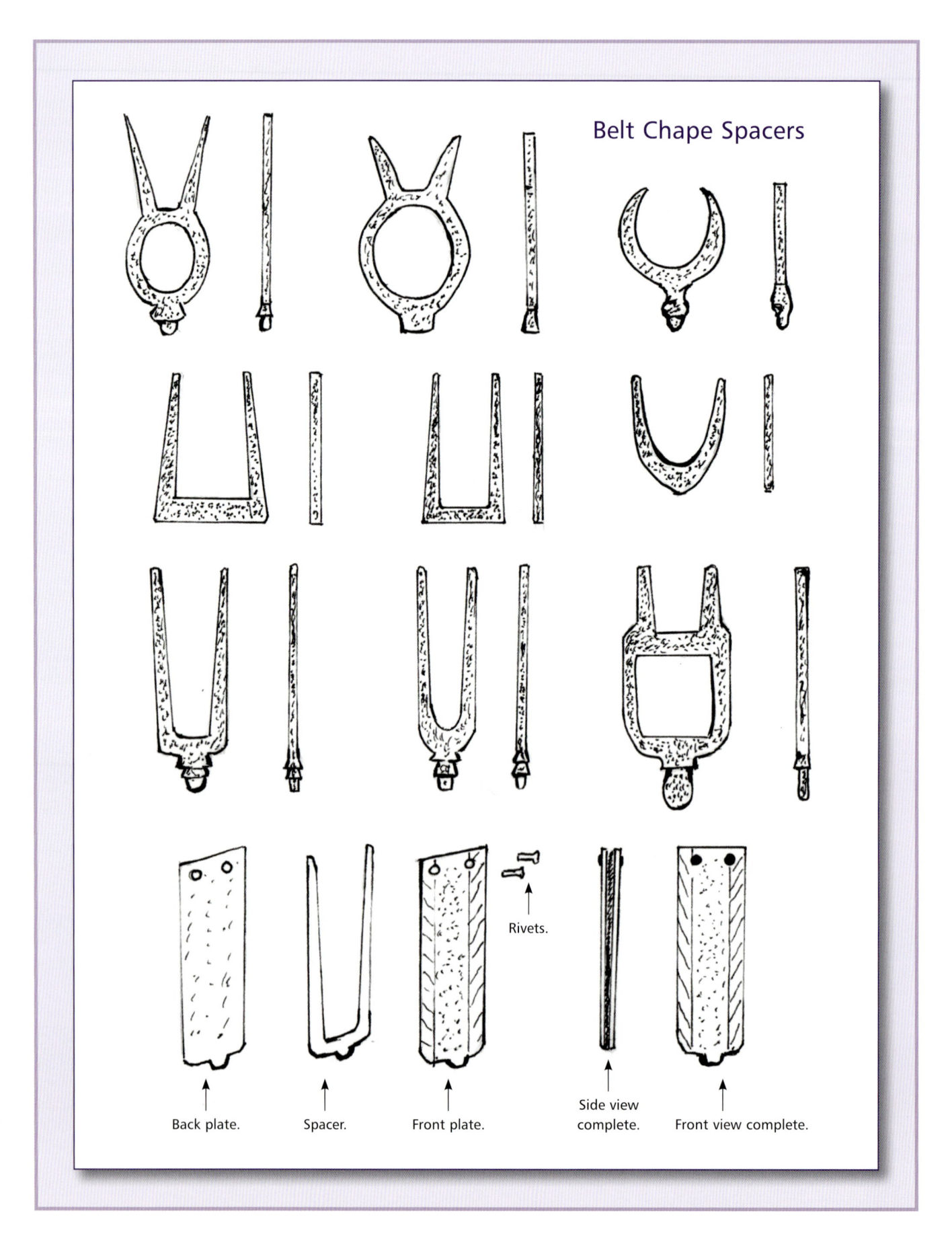

Belt Chapes

Belt Chape Spacers

Rivets.

Back plate.   Spacer.   Front plate.   Side view complete.   Front view complete.

Belt Chapes – Roman

as items of fashion, they were meant to be seen and admired. The width of the belt was not always reflected in the size of the chape. Some belts would have been of the same width as the chape, but others wider or ever narrower.

Many of these early chapes are recovered in areas where the military were stationed, camped, travelled or fought. Belt decorations such as the chape or stiffeners would have been more frequently lost than buckles during this period. However, when one approaches the medieval period this is reversed and buckles become the more common find.

With the departure of the Romans the Jutes, Angles and Saxons were the next to settle in this land and, as with the Romans, they brought their own culture and fashion.

Fig.1. Roman amphora-shaped chape the handles supported by two square pillars decorated with rings and dots. The body has a star surrounded by small dots and circles. This is a large example, and dates 4th century.

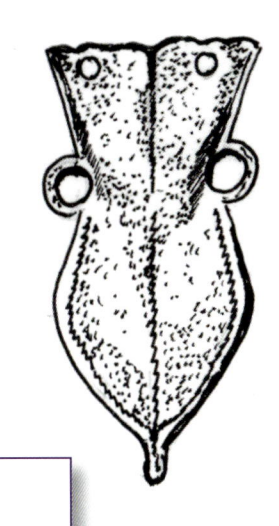

Fig.2. Roman amphora-shaped chape with round handles and a wriggle work border, 4th century.

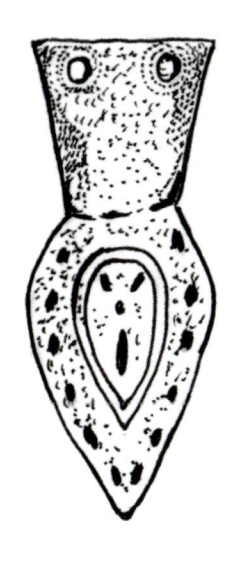

Fig.3. Roman amphora-shaped chape, no handles, shield-shaped centre design with three small lines and a central dot; the whole surrounded by an outer border of dots. It would date 4th-5th century.

Fig.4. Roman amphora (almost spear-shaped) chape, no handles, foliage design; 5th century in date.

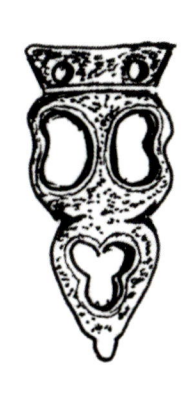

Fig.5. Roman amphora-shaped chape, the inner open handles are kidney shaped the lower part open trefoil design, 5th century.

67

Belt Chapes – Roman

Fig.6. Roman amphora shape chape no handles the central design a large letter "S" surrounded by a rope border, 5th century.

Fig.7. Roman amphora-shaped chape inner open handles are kidney shaped, crude triple lined circle on the body, 5th century.

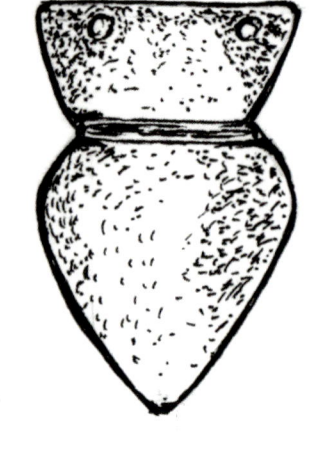

Fig.8. Roman amphora-shaped chape, plain body, no handles, 5th century.

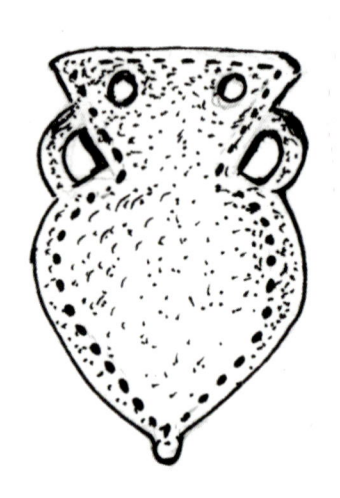

Fig.9. Roman amphora-shaped chape with handles, outer border decorated with small dots, 4th century.

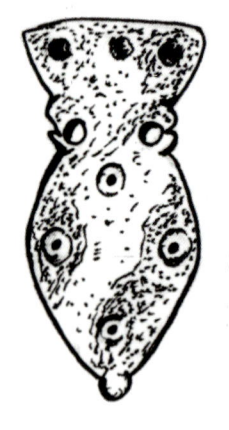

Fig.10. Roman amphora-shaped chape, two small holes that could represent handles; the body decorated with four rings and dots, 5th century.

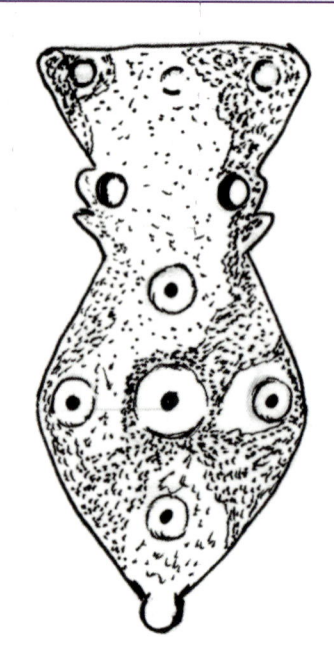

Fig.11. Roman amphora-shaped chape; the two small holes could represent the handles. The body has a central open design with three rings and dots; this is a large example and dates 4th century.

Belt Chapes – Roman

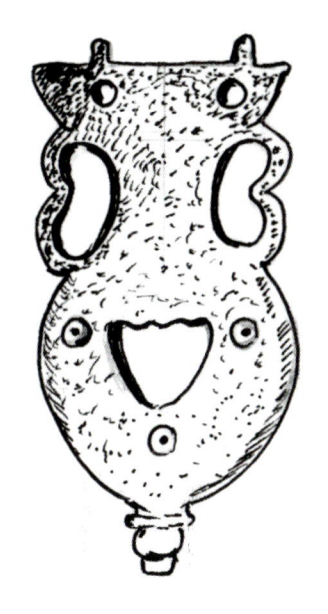

Fig.12. Roman amphora-shaped chape the handles are kidney shaped while the body has an open design with three rings and dots; 4th century.

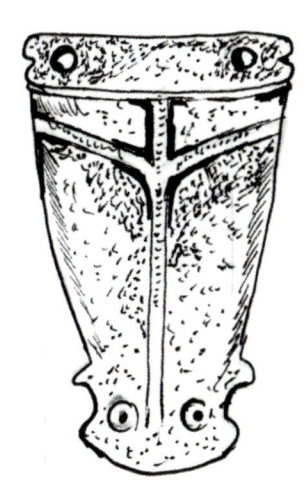

Fig.13. Roman shield-shaped chape with a lined design almost in the form of a cross with two rings and dots; 4th century.

Fig.14. Roman amphora-shaped chape with "D" shaped handles terminating in loops. The body has a wriggle work border with eight rings and dots; 4th century.

Fig.15. Roman amphora-shaped chape; no handles but an inner triple circle design with a radiate ring and dot. The top part has a double rope design; 4th century.

69

Belt Chapes – Roman

Fig.16. Roman amphora oblong-shaped chape; small handles, one ring and dot. This dates 5th century.

Fig.17. Roman oblong-shaped chape terminating in a fork design; it has one ring and dot, and dates 5th century.

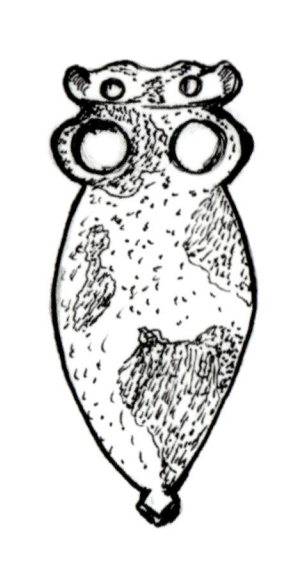

Fig.18. Roman amphora-shaped chape with plain body, 4th century.

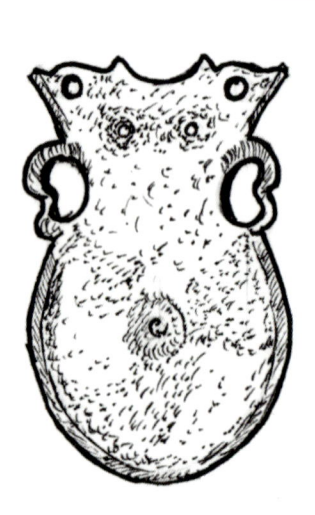

Fig.19. Roman amphora-shaped chape, kidney-shaped handles, scalloped top, single ring and dot, 4th century.

Pic.1. Roman amphora-shaped chape, the handles supported by two square pillars decorated with rings and dots. The body has a star surrounded by small dots and circles. This is a large example, 4th century.

Belt Chapes – Saxon

Though the Saxons settled in areas of Essex, Sussex and Wessex the Jutes settled in Kent and the Isle of White, while the Angles controlled East Anglia, Northumbria and Mercia.

This was not the first time that the invaders had set foot on our shores for they had carried out raids from the late 3rd century into and throughout the 4th century. The chapes that adorned their belts differed greatly from their Celtic and Roman predecessors as they were normally much smaller and more highly decorated. They were usually made from bronze,

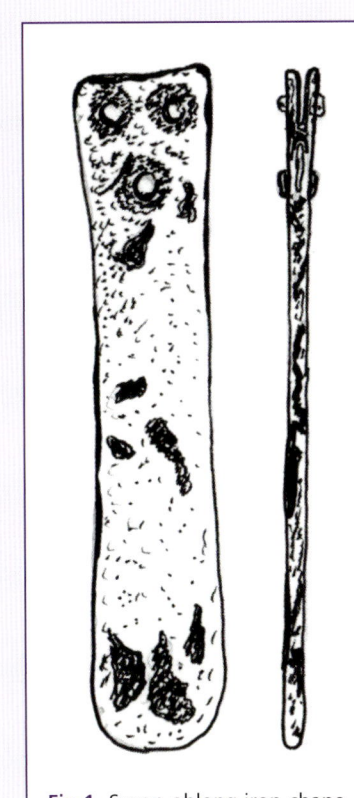

Fig.1. Saxon oblong iron chape, plain body three rivets, late Saxon.

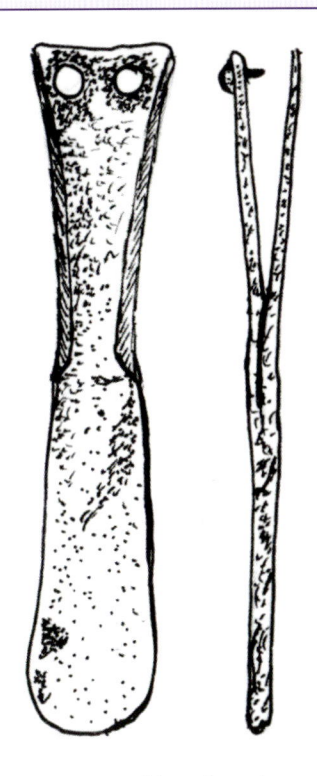

Fig.2. Saxon oblong iron chape, chamfered sides at top half, plain body, two rivets, late Saxon.

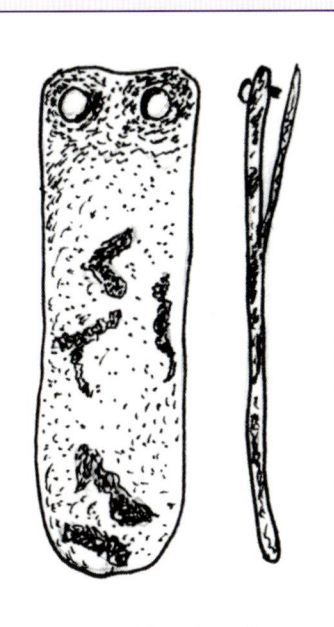

Fig.3. Saxon oblong iron chape, plain body, two rivets, late Saxon.

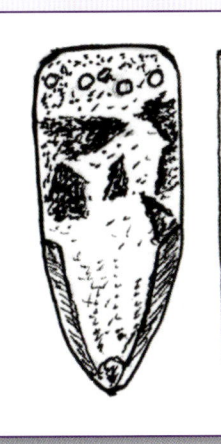

Fig.4. Saxon shield-shaped iron chape, plain body, four rivet holes, late Saxon.

Fig.5. Saxon bronze chape, animal facing upwards within a roped border, two rivet holes, 9th century.

Fig.6. Saxon bronze chape, interlaced design, beaded border, two rivets, 9th century.

71

Belt Chapes – Saxon

Fig.7. Saxon bronze chape, interlaced design, two rivet holes, 9th century.

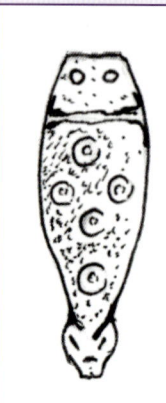

Fig.8. Saxon bronze chape, ring and dot design, two rivet holes, 9th century.

Fig.9. Saxon bronze chape, interlaced design, two rivet holes, 9th century.

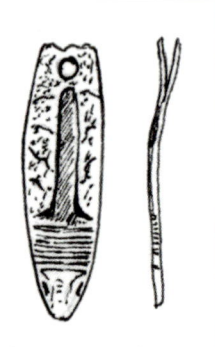

Fig.10. Saxon bronze chape, banded with vertical design, one rivet hole, 9th century.

Fig.11. Saxon bronze chape, multiple interlock animals, two rivet holes, 9th century.

Fig.12. Saxon bronze chape, two animals within a border, two rivet holes, 9th century.

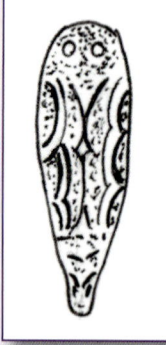

Fig.13. Saxon bronze chape, mirrored crescent design, two rivet holes, 9th century.

Fig.14. Saxon bronze chape, interlaced design, two rivet holes, 9th century.

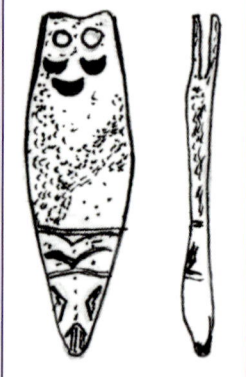

Fig.15. Saxon bronze chape, three crescents, two rivet holes, 9th century.

Fig.16. Saxon bronze chape, unusual shape, interwoven lines, two rivets, 9th century.

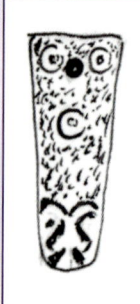

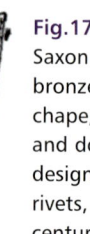

Fig.17. Saxon bronze chape, ring and dot design, two rivets, 9th century.

Belt Chapes – Saxon

with some examples inlaid with silver or gold.

Some chapes manufactured from pure silver or gold have also been recovered, but are extremely rare. Iron chapes were also made and used, but due to the rapid deterioration of this metal in the ground few have survived.

The majority of surviving Anglo-Saxon chapes are from the 9th century, and turn up either as casual losses or as grave goods.

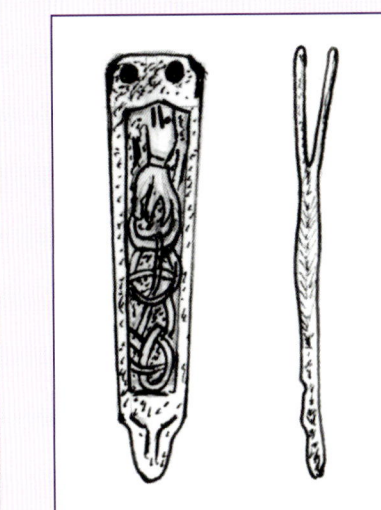

Fig.18. Saxon bronze chape, howling beast with interwoven design below two rivet holes, 9th century.

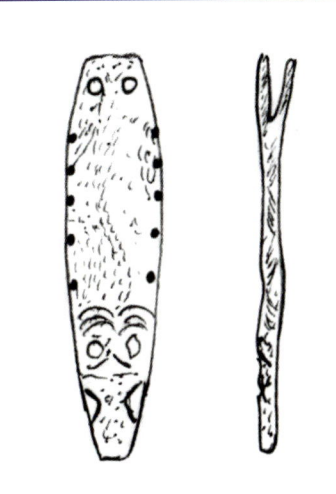

Fig.19. Saxon bronze chape, dotted border below two dots with crescent design, two rivet holes, 9th century.

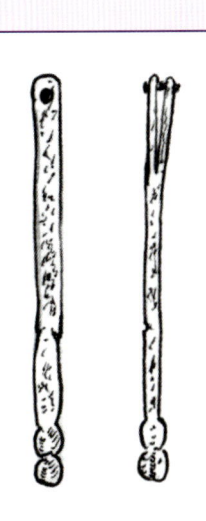

Fig.20. Saxon bronze oblong thin chape, terminating in a double knop, one rivet, 9th-10th century.

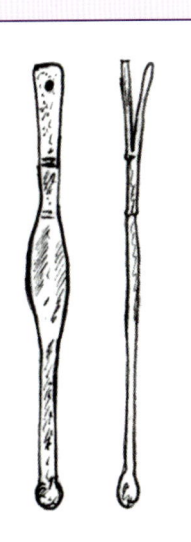

Fig.21. Saxon bronze oblong chape, the centre lozenge shaped, one rivet hole, 9th-10th century.

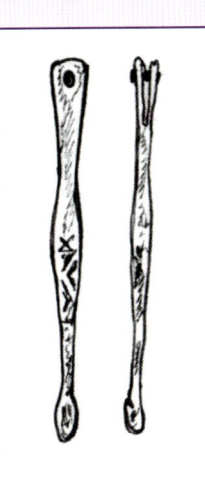

Fig.22. Saxon bronze oblong chape, bulbous centre, interlaced design, one rivet, 9th-10th century.

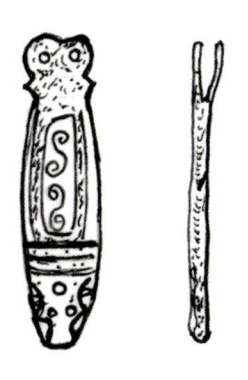

Fig.23. Saxon bronze chape, scrolled design within a double line, two rivet holes, 9th century.

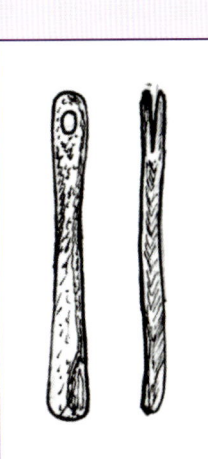

Fig.24. Saxon bronze oblong chape, one rivet hole, 9th century.

Belt Chapes – Saxon

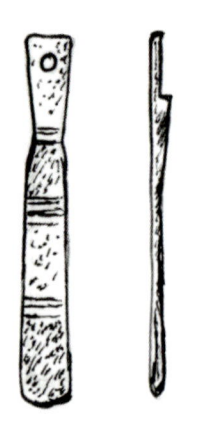

Fig.25. Saxon bronze oblong chape, triple banded design, one rivet hole 9th-10th century.

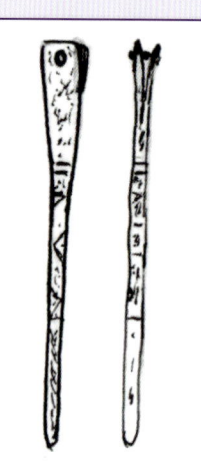

Fig.26. Saxon bronze chape, lined design within a beaded border, one rivet, 9th century.

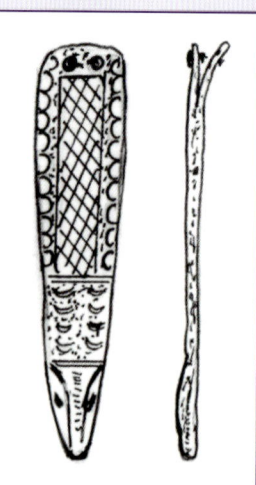

Fig.27. Saxon bronze chape, cross-hatched design within a beaded border, two rivets, 8th-10th century.

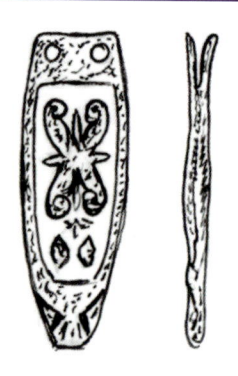

Fig.28. Saxon bronze chape, almost a mirrored fleur-de-lis design within a single lined border, two rivet holes, 9th century.

Fig.29. Saxon bronze chape, mirrored double-arch design with triple-roped band below two rivet holes, 9th century.

Fig.30. Saxon bronze chape terminating in two raised bands, with two squares and two circles above one rivet hole, 9th century.

Fig.31. Saxon bronze chape, animal mask type, two rivet holes, 9th century.

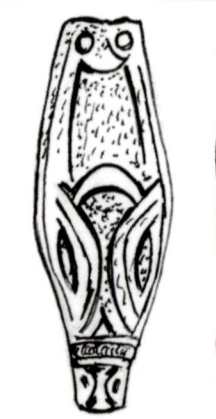

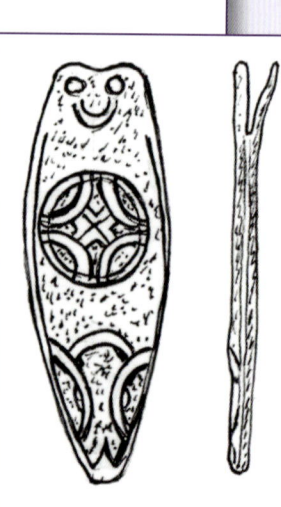

Fig.32. Saxon bronze chape, arc type, two rivet holes, 9th century.

Fig.33. Saxon bronze chape, arc type, two rivet holes, 9th century.

Belt Chapes – Saxon

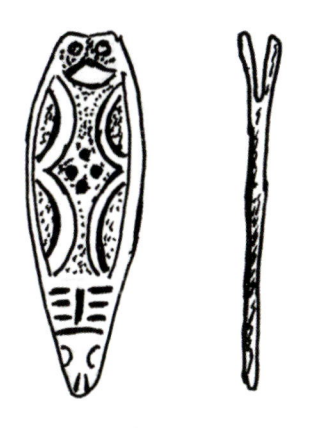

Fig.34. Saxon bronze chape, arc type, two rivet holes, 9th century.

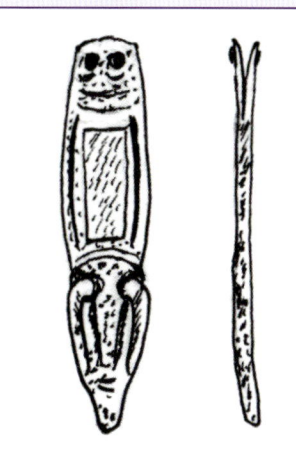

Fig.35. Saxon bronze chape, animal mask type, two rivet holes, 9th century.

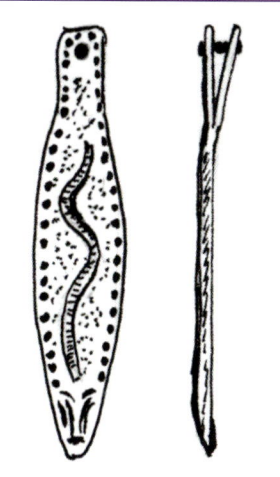

Fig.36. Saxon bronze chape, central scroll surrounded by a beaded border, one rivet, 9th century.

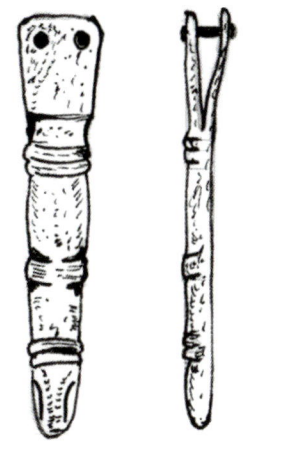

Fig.37. Saxon bronze chape, banded type, two rivets, 9th century.

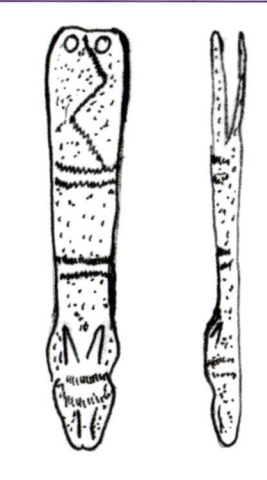

Fig.38. Saxon bronze chape, banded type, two rivet holes, 9th century.

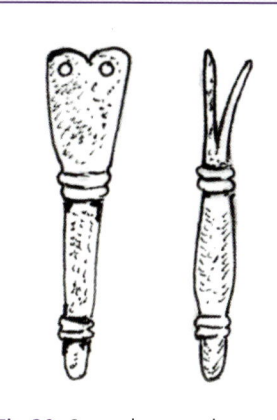

Fig.39. Saxon bronze chape, banded type, thinner lower half, two rivet holes, 9th century.

Fig.40. Saxon bronze chape, interlaced design, two rivets, 9th century.

Fig.41. Saxon bronze chape, banded type, thin body, two rivets, 9th-10th century.

Fig.42. Saxon bronze chape, banded type, thin body, two rivets, 9th-10th century.

75

Belt Chapes – Saxon

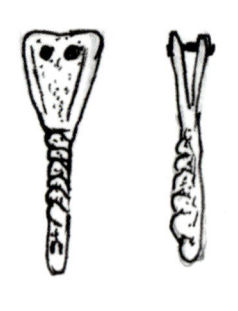

Fig.43. Saxon bronze chape, thin spiral body, two rivets, 9th century.

Fig.44. Saxon bronze chape, roped border, two rivets, 9th century.

Fig.45. Saxon bronze chape, open centre surrounded by a foliage design, two rivets, 9th century.

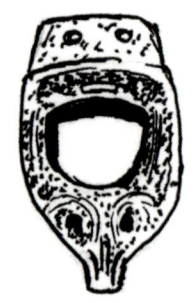

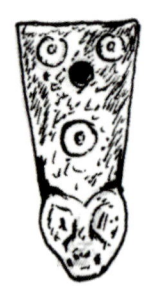

Fig.46. Saxon bronze chape, ring and dot design, one rivet, 9th-10th century.

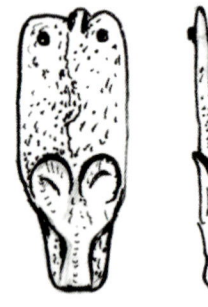

Fig.47. Saxon bronze chape, wolf head design, two rivet holes, 9th century.

Fig.48. Saxon bronze chape, raised bands with ring and dot design, two rivets, 9th century.

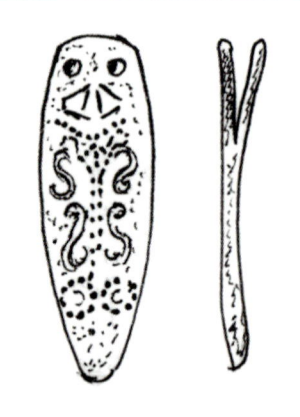

Fig.49. Saxon bronze chape, scroll design, two rivet holes, 9th century.

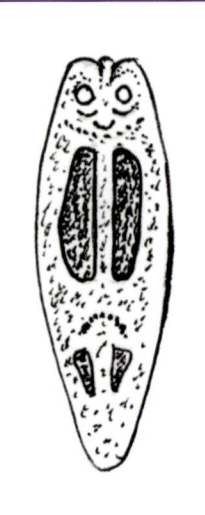

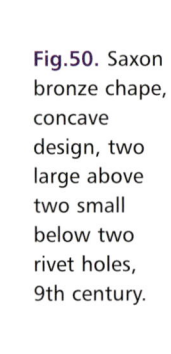

Fig.50. Saxon bronze chape, concave design, two large above two small below two rivet holes, 9th century.

Belt Chapes – Saxon

Pic.1. Late Saxon shield-shaped iron chape, plain body, two rivets.

Pic.2. Late Saxon oblong iron chape, chamfered sides at top half, plain body, two rivets.

Pic.3. Late Saxon oblong iron chape, plain body, two rivets.

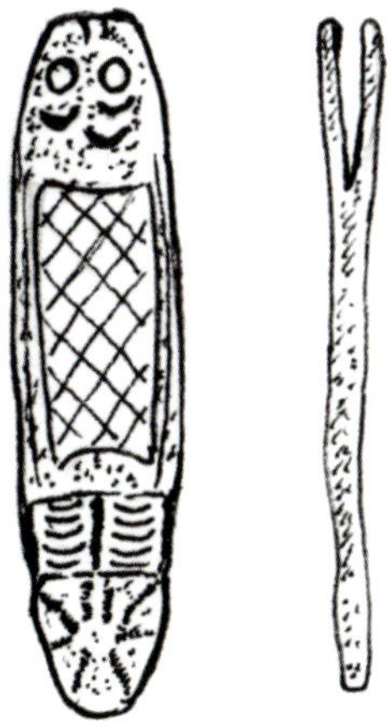

Fig.51. Saxon bronze chape, cross hatch within a lined border, two rivet holes, 8th-10th century.

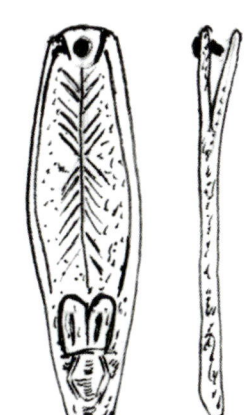

Pics.6. Saxon bronze chape, multiple interlock animals, two rivet holes, 9th century.

Pic.5. Saxon bronze chape, banded with vertical design, one rivet hole, 9th century.

Fig.52. Saxon bronze chape, herring bone design within a lined border, 9th century.

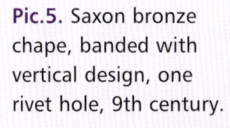

Pic.4. Late Saxon oblong iron chape, plain body, three rivets.

Pic.7. Saxon bronze chape interlaced design two rivet holes 9th century.

77

Belt Chapes – Viking

The Vikings are always thought of mainly as raiders and traders, which is correct. However, after the middle of the 9th century they arrived to settle in large numbers in eastern and northern England. Their main aim was to conquer and colonise England, and between the years 865-878 they destroyed the ancient Anglo-Saxon kingdoms of East Anglia and Northumbria and went on to conquer part of Mercia.

They were eventually defeated in AD 878 at the battle of Eddington, which resulted in a boundary between Anglo-Saxon Wessex and Viking East Anglia – the division boundary being the old Roman road of Watling Street between London and Chester.

The Vikings, like the previous settlers, brought with them their own

Fig.1. Viking bronze chape beaded border two double lined diamonds with pellets two rivet holes.

Fig.2. Viking bronze chape, ringed chain/bow design terminating in an animal mask, two rivet holes.

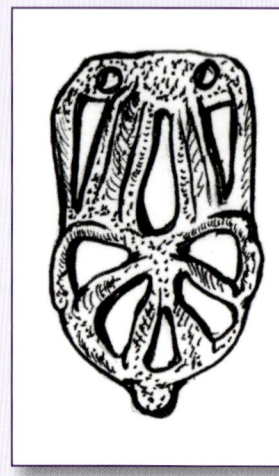

Fig.3. Viking bronze chape, open work design, two rivet holes.

Fig.4. Viking bronze chape, ring/chain design, two rivet holes.

Fig.5. Saxon/Viking bronze chape in the form of a wolf's head, two rivet holes.

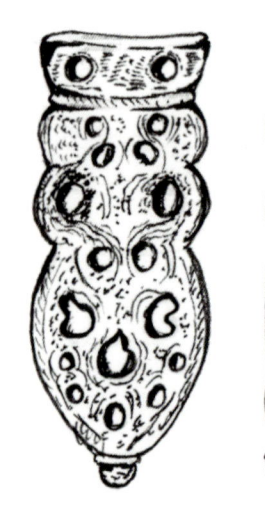

Fig.6. Viking bronze chape, amphora shape with open work, two rivet holes.

Belt Chapes – Viking

culture, language and fashion, and the design of their metalwork obviously differed from that of the other settlers. Their chapes are generally much heavier, robust and larger.

In most cases they are also extremely well executed; however, over a period of time copies would be produced from the original. Eventually, after copy from copy the sharpness of detail would deteriorate and the chapes would appear to be of poor quality.

Viking chapes were normally decorated only on one side. However, a group of known examples exist carrying a design on both sides. This decoration consisted of interlace or knot-work pattern, with the terminal itself often having a double-lined wheel with a pierced centre.

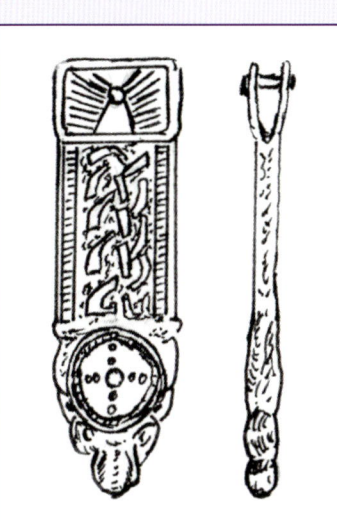

Fig.7. Viking bronze chape, interlaced design between two columns, all below a beaded wheel, one rivet.

Fig.8. Viking bronze chape, crude irregular design, two rivet holes.

Fig.9. Viking bronze chape, three protruding lugs with round concave inserts below a roped horizontal line, three rivet holes.

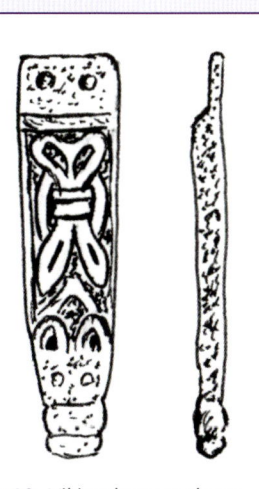

Fig.10. Viking bronze chape, ringed chain/bow design terminating in a animal mask, two rivet holes.

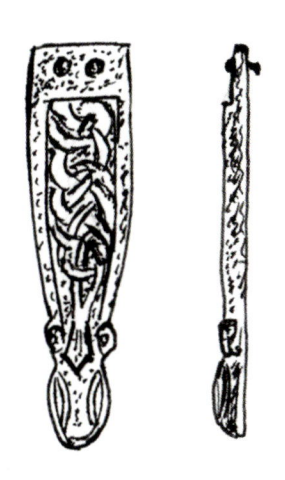

Fig.11. Viking bronze chape, ring chain within a single lined border, two rivets.

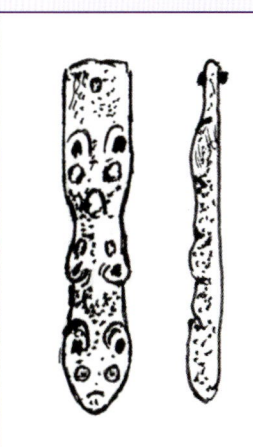

Fig.12. Viking bronze chape, triple animal head, one rivet.

79

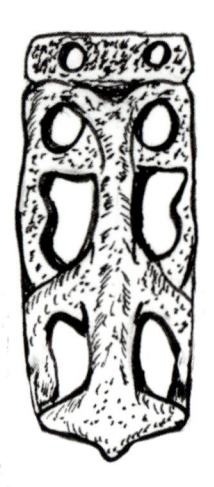

Belt Chapes – Viking

Fig.13. Viking bronze chape, crude openwork design, two rivet holes.

Fig.14. Viking bronze chape, ring chain design with a form of human face, two rivet holes.

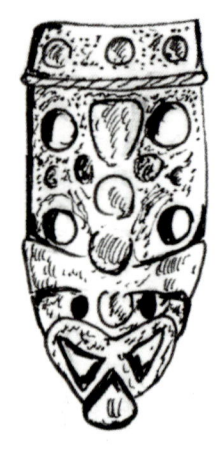

Fig.15. Viking bronze chape, horizontal roped design at top; open worked with four raised bosses terminating in the form of a mask, three rivet holes.

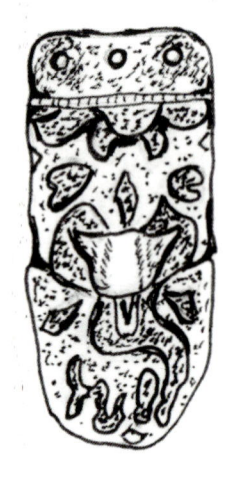

Fig.16. Viking bronze chape, roped horizontal design at top below irregular design almost representing falling leaves, three rivet holes.

Fig.17. Viking bronze chape, four equal squares each having a ring and dot with radiate lines, the whole being surrounded by a ring and dot border, the lower examples divided by double lines; two rivet holes.

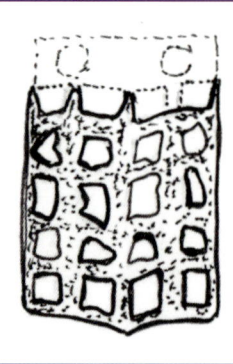

Fig.18. Viking bronze chape (top rivet section missing), 16 irregular openwork squares; possibly two rivet holes.

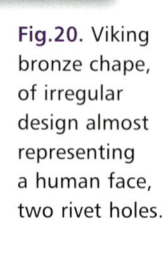

Fig.19. Viking bronze chape, unusually-shaped irregular design, slit at top for leather strap, two holes although somewhat large for rivets.

Fig.20. Viking bronze chape, of irregular design almost representing a human face, two rivet holes.

Fig.21. Viking bronze chape, irregular radiate lines, two rivet holes.

80

Belt Chapes – Viking

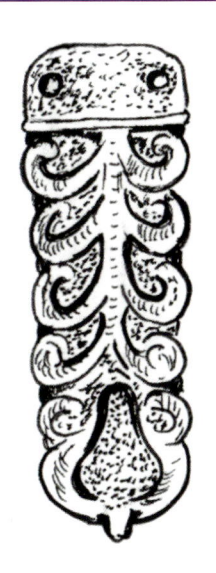

Fig.22. Viking bronze chape representing a tree terminating in almost a tear drop, two rivet holes.

Fig.23. Viking bronze chape, very slender in shape, one rivet hole.

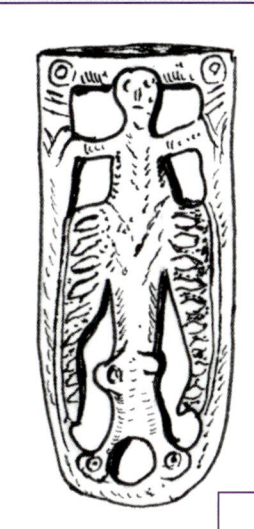

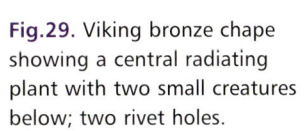

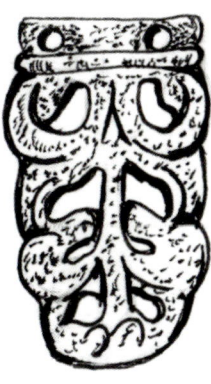

Fig.24. Viking bronze chape, openwork design in the form of a winged creature or man of unusual proportions, two rivet holes.

Fig.25. Viking bronze chape having the face of a bearded man, two rivet holes.

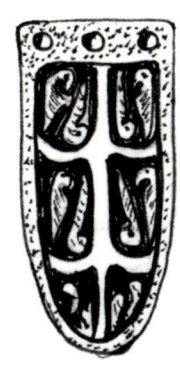

Fig.26. Viking bronze chape, shield-shaped design divided into six parts, each having an oblong scroll; three rivet holes.

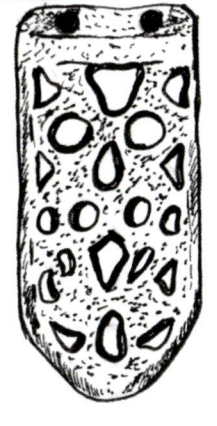

Fig.27. Viking bronze chape, crude irregular openwork design consisting of 20 holes; two rivets.

Fig.28. Viking bronze chape showing a bearded man with raised arms and below which appears to be two mirrored creatures; two rivet holes.

Fig.29. Viking bronze chape showing a central radiating plant with two small creatures below; two rivet holes.

Belt Chapes – Viking

Fig.30. Viking bronze chape, openwork design below a horizontal rope. At the top numerous rings and dots; two rivet holes.

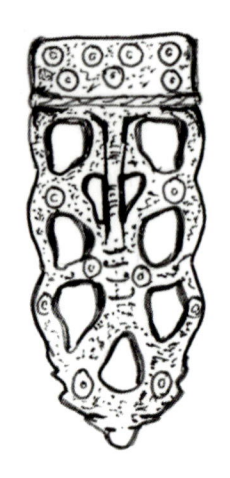

Fig.31. Viking bronze chape, openwork below a horizontal roped design with ring and dot above; two rivet holes.

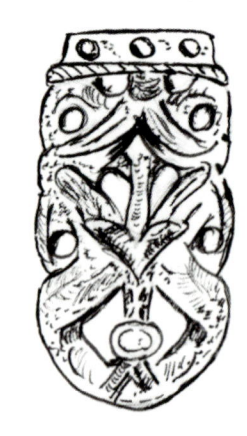

Pic.1. Viking bronze chape (top rivet section missing), 16 irregular openwork squares, possibly two rivet holes.

Fig.34. Viking bronze chape, some openwork with raised foliage design radiating to the edges below a horizontal rope, three rivet holes.

Fig.32. Viking bronze chape, openwork design below a horizontal rope; two rivet holes.

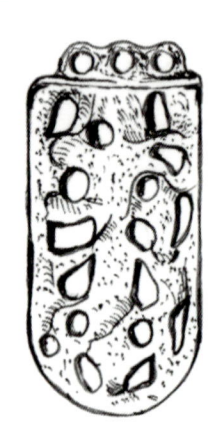

Fig.33. Viking bronze chape, open design, three rivet holes.

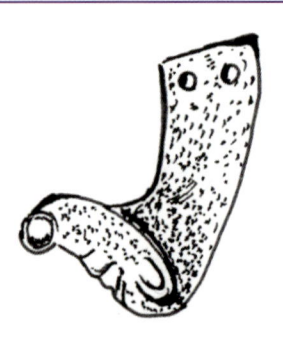

Fig.35. Viking bronze chape terminating in a mythical beast in the Ringerike style, two rivet holes.

Belt Chapes – Viking

Pic.3. Viking bronze chape, unusually shaped irregular design, slit at top for leather strap, two holes though somewhat large for rivets.

Pics.2a & b. Saxon/Viking bronze chape in the form of a wolf's head, two rivet holes.

Fig.36. Viking bronze chape, openwork that could represent a tree below a horizontal rope, three rivet holes.

Pic.4. Viking bronze chape, central boss surrounded by repeated radiate design, three rivet holes.

Fig.37. Viking bronze chape, rather worn design having two raised central bosses surrounded by crescents around the edge below a horizontal rope, two rivet holes.

Belt Chapes – Norman

The Normans were next to conquer this land, once again bringing with them their own laws, culture and fashions. As might be expected their style of belt decorations also differed greatly. Chapes from this period are rare, unlike those of their predecessors which appear to have been more plentiful (even including those from the Viking period).

There are no particular styles that appear to have been more favoured than others, although the crucifix seems to have been popular. Many Norman chapes were solid cast and of good workmanship. To date I have not seen a poor copy from an original, though they could well exist.

Fig.1. Norman bronze chape, openwork below a triple band, two rivet holes.

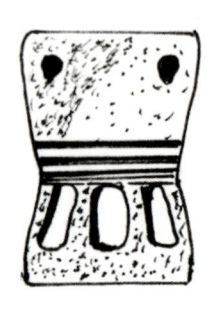

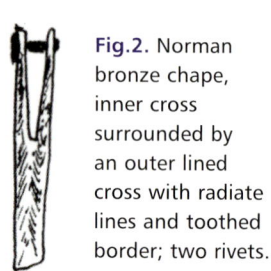

Fig.2. Norman bronze chape, inner cross surrounded by an outer lined cross with radiate lines and toothed border; two rivets.

Fig.3. Norman bronze chape, central triple diamond with double lined "V" above and below, the whole surrounded by a lined border; two rivet holes.

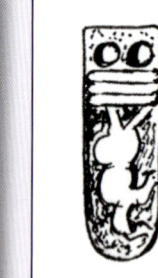

Fig.4. Norman bronze chape, possibly a mythical creature below a triple band; two rivet holes.

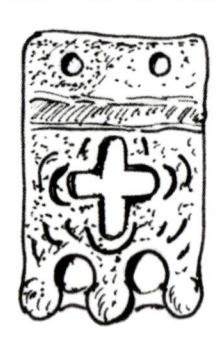

Fig.5. Norman bronze chape, openwork cross with two lower holes beneath a double band terminating in triple knops with two central bars; two rivets.

Fig.6. Norman bronze chape, lower part having four holes between a double lined border, above two holes with a triangular hole. The whole of the body decorated in irregular double lines and terminating in a knop; two rivet holes.

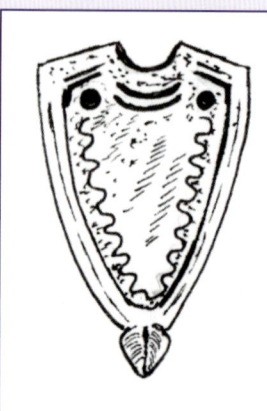

Fig.7. Norman bronze chape, shield shape terminating in a knop; wriggle work design with two crescents above surrounded by a double lined border; two rivets.

Belt Chapes – Norman

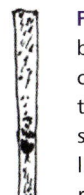

Fig.8. Norman bronze chape, crescent shape, top wriggle work surrounded by a lined border; two rivet holes.

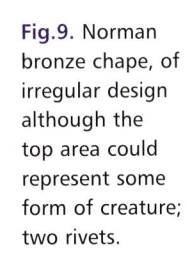

Fig.9. Norman bronze chape, of irregular design although the top area could represent some form of creature; two rivets.

Fig.10. Norman bronze chape, fleur-de-lis within a mirrored "V", cross-hatched background; two rivet holes.

Fig.11. Norman bronze chape showing a mythical winged creature facing backward within a lined border; two rivet holes.

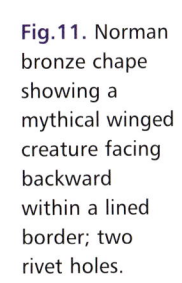

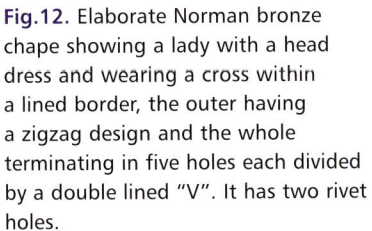

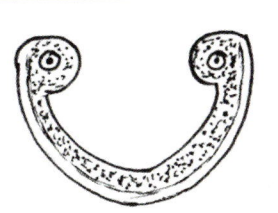

Fig.12. Elaborate Norman bronze chape showing a lady with a head dress and wearing a cross within a lined border, the outer having a zigzag design and the whole terminating in five holes each divided by a double lined "V". It has two rivet holes.

Fig.13. Norman bronze chape, crescent shaped, lined border, two rivet holes.

Pic.1. Norman bronze pear-shaped chape terminating in a "V", the body having 17 pellets within a lined border; one rivet hole.

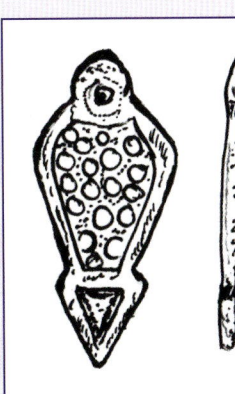

Fig.14. Norman bronze chape, pear shaped, terminating in a "V" and the body having 17 pellets within a lined border; one rivet.

85

Belt Chapes – Medieval

Most of the chapes recovered in Britain date from the 12th-16th centuries, this being the period when they were widely used by both men and women. As with many everyday items the chape was made from various metals including bronze, copper (normally gilded, silvered or tinned) pewter, tin, silver and gold, and iron (that was normally coated to prevent rust).

The chapes made from precious metal would normally have been melted down into items more fashionable when they fell from favour. This is why few have survived, unless they had been lost in the ground. The workmanship lavished on some of these chapes was of a standard that would have graced the most expensive jewellery.

Some solid silver chapes have been recovered, though few in number. The examples I have seen, although not large in size, were fine pieces of medieval jewellery of outstanding workmanship.

Two examples of these chapes I have encountered would have once contained a religious relic. The first example had a body that was hinged. It was decorated with a floral spray and terminated in a knop. The other, with a body that was not hinged, showed

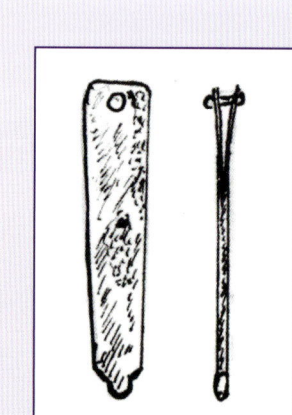

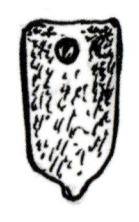

Fig.1. Medieval copper-alloy chape, plain body, one rivet.

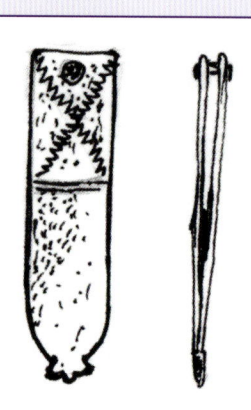

Fig.2. Medieval copper-alloy chape terminating in a trefoil, crossed wriggle work above a double line, single rivet.

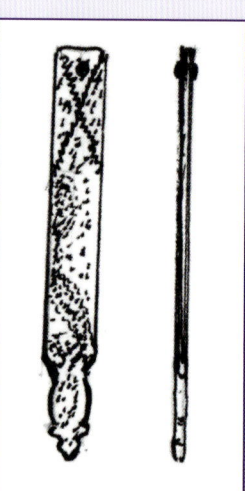

Fig.3. Medieval copper-alloy chape terminating in a trefoil, crossed wriggle work, one rivet.

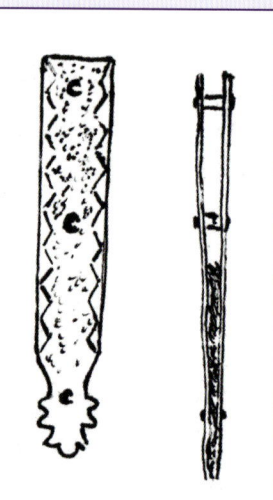

Fig.4. Medieval copper-alloy chape terminating in seven radiate points, zigzag decoration, two rivets.

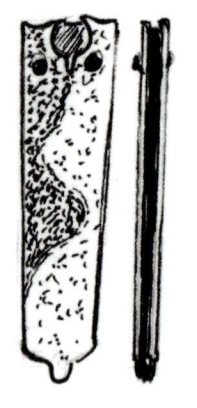

Fig.5. Medieval copper-alloy chape, crescent top, plain body, two rivets.

Fig.6. Medieval copper-alloy chape, plain body, one rivet.

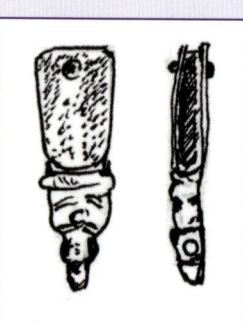

Fig.7. Medieval copper-alloy chape terminating in an animal head, plain body, one rivet.

Fig.8. Medieval copper-alloy chape, plain body, two rivets.

Belt Chapes – Medieval

a figure of a saint on the obverse side with the reverse being plain. Although both of these chapes were opened, they were then carefully resealed without disturbing the contents.

Occasionally, some of the leather or other original material from the belt remains intact inside the chape, thereby adding to its interest.

Many medieval paintings, brasses and effigies show the wealthy in all their finery. Costumes are shown in great detail and the girdle (belt) was often enriched leather adorned with an ornate buckle and, of course, a chape.

The one-piece cast chapes or strap ends were widely used until the 13th century and then fell from favour for some years, regaining popularity in the late 14th century. There is a great variety in the sizes of chapes from this period although they are usually rectangular in shape. Even the small cast chapes are quite heavy, and this may well have been the intention so that the weight kept the excess of the belt or girdle hanging straight down.

Chapes were not only used on girdles or belts; any type of leather strap could have been fitted with them. Other types of chapes consisted of a single piece of metal that was folded into a U-shape and secured to the belt

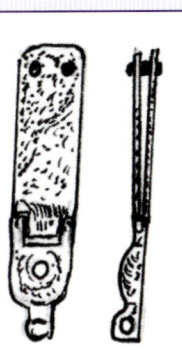

Fig.9. Medieval copper-alloy chape, hinged plate and loop, plain body, two rivets.

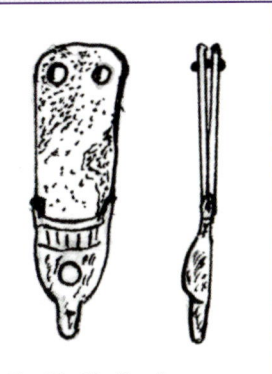

Fig.10. Medieval copper-alloy chape, hinged plate and loop, plain body, two rivets.

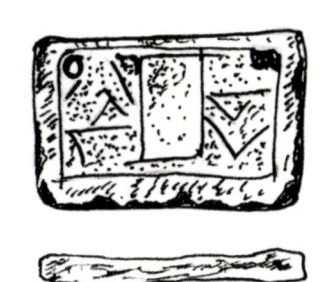

Fig.11. Medieval copper-alloy chape, folded sheet with turned ends, geometric design, two rivet holes.

Fig.12. Medieval copper-alloy chape folded sheet open sides, raised grooved one rivet.

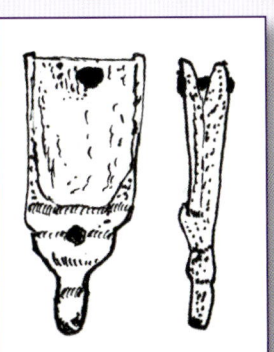

Fig.13. Medieval copper-alloy chape terminating in an animal head, one rivet.

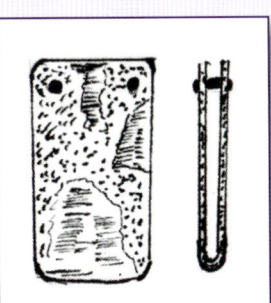

Fig.14. Medieval copper-alloy chape, folded sheet with open sides, plain body, two rivets.

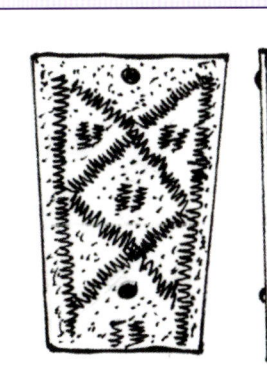

Fig.15. Medieval copper-alloy chape engraved with two wriggle work "Xs" surrounded by a border, two rivets.

Belt Chapes – Medieval

by one or two rivets. Another form would, as before, have been made from a rectangular piece of sheet metal, folded on itself into a U-shape. A thin piece of strip metal would then have been folded into a "U" and soldered onto the first piece to form the sides. The leather strap would have been held in place by rivets.

Two-piece chapes could best be described as "double sheet", this being two sheets of metal attached to the leather strap by means of rivets.

A second type of two-piece chape incorporated bent slides on one of the plates, thereby concealing and protecting the edge of the leather strap. The third type of two-piece chape had a cast front plate, but a simple metal sheet back. The cast section – as with the one-piece types – would normally have been ornate. The reverse plate would have been concave in section to accommodate the leather strap. Usually two securing rivets, at the top and bottom, would have held the plates together.

Three-piece chapes were made in large numbers, but are most common in the 15th-16th centuries. Should these chapes carry decoration (for in the main they are normally plain) this is usually on one side. The style of decoration can vary from simple zigzag lines to intricate patterns. Besides such decoration, fancy knops were also used on the part of the spacers protruding from the chape.

The most favoured knop was that of the acorn, with others being: collared or grooved, trefoil, lozenge, or tongue shape. Multi-shaped other

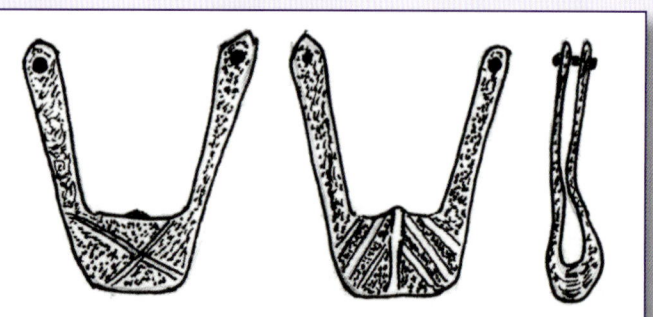

Fig.16. Medieval copper-alloy chape, folded sheet, "V" shaped one side having a double lined "X" the other, herring bone with central line, two rivets.

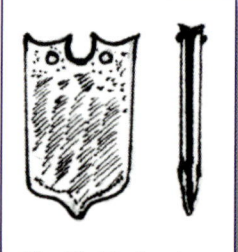

Fig.17. Medieval copper-alloy chape, crescent-shaped top, plain body, two rivets.

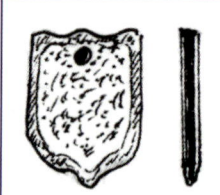

Fig.18. Medieval copper-alloy chape, shield shaped, with lined border, one rivet.

Fig.19. Medieval copper-alloy chape, shield shaped with cross hatching, two rivets.

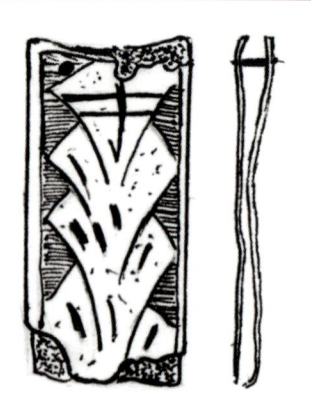

Fig.20. Medieval copper chape decorated with interlocking crude shaped chevrons, two rivets.

Fig.21. Medieval copper-alloy chape, pear shaped, quatrefoil in centre with cross hatching above, surrounded by a lined border, two rivets.

Belt Chapes – Medieval

Fig.22. Medieval copper-alloy chape, lower half having 11 radiate lobes above open kidney shape with geometric double lines, two rivets.

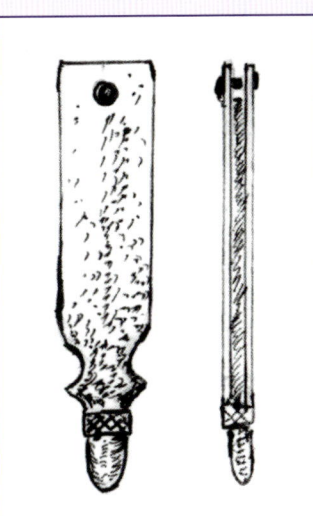

Fig.23. Medieval copper-alloy chape terminating in crescent-shaped notches each side with a acorn below, one rivet.

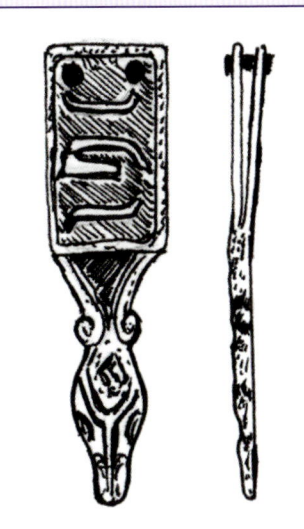

Fig.24. Medieval copper-alloy chape three black letters within a lined border terminating in a animal head, two rivets.

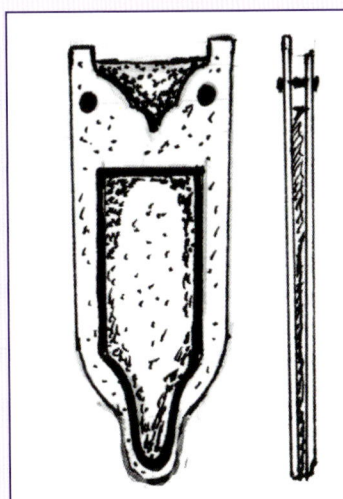

Fig.25. Copper-alloy chape, scalloped top, tongue-shaped design, two rivets.

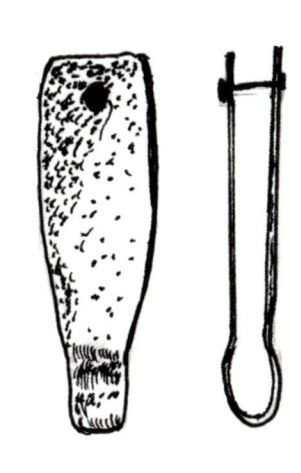

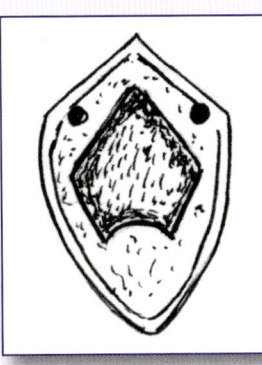

Fig.26. Medieval copper-alloy chape, folded sheet, open sides, plain body, two rivets.

Fig.27. Medieval copper-alloy chape, folded sheet, open sides, geometric design surrounded by a lined border, one rivet.

Fig.28. Medieval copper-alloy chape, shield-shaped bishop mitre design surrounded by a lined border, one rivet.

Belt Chapes – Medieval

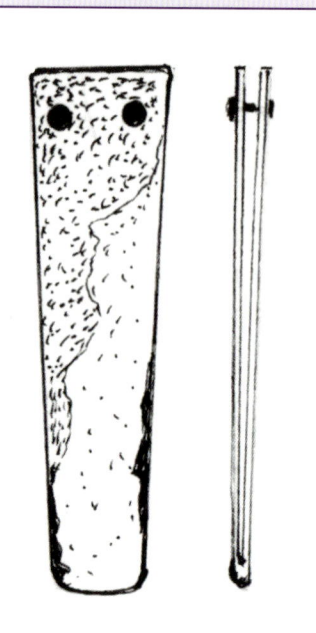

Fig.29. Medieval copper-alloy chape, folded sheet, plain body, two rivets.

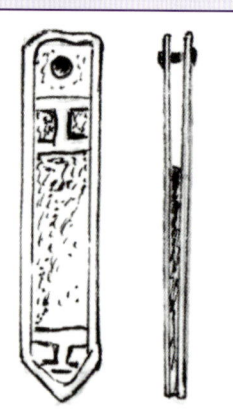

Fig.30. Medieval copper-alloy chape, two horizontal "Hs" surrounded by a lined border, one rivet.

Fig.31. Medieval copper-alloy chape, plain body, one large dome-shaped rivet.

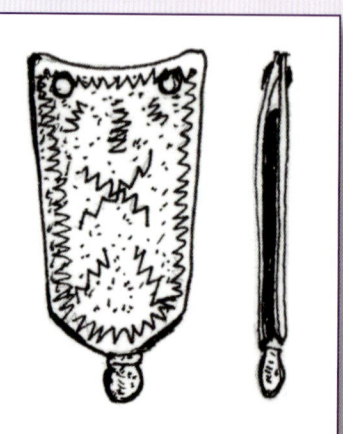

Fig.32. Medieval copper-alloy chape terminating in a knop, crude wriggle work "X" surrounded by a border, two rivets.

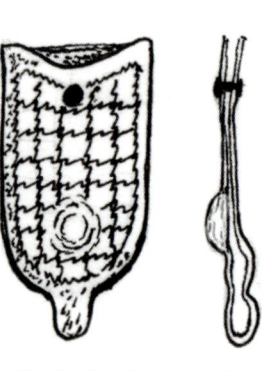

Fig.33. Medieval copper-alloy chape, folded sheet, open sides terminating in a open knop, raised boss surrounded by wriggle work in the form of squares, one rivet.

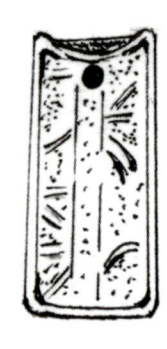

Fig.34. Medieval copper-alloy chape, folded sheet, open sides geometric design, one rivet.

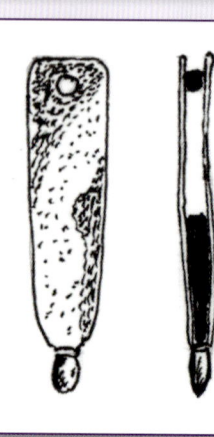

Fig.35. Medieval copper-alloy chape terminating in knop, plain body, one rivet.

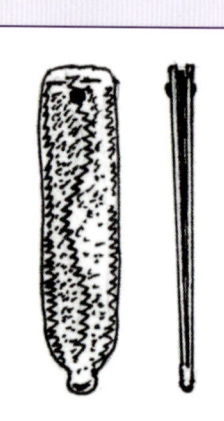

Fig.36. Medieval copper-alloy chape terminating in a knop, wriggle work oblong scroll surrounded by a border, one rivet.

Belt Chapes – Medieval

types are known, but the ones previously mentioned were the most widely used.

There is another type of chape that is often described as for use on the belt or girdle; however, this falls into a grey area for no one is sure as to its true use. This chape is a two-piece – the parts being held together by a single rivet. The lower half is cast, normally carries some decoration, and terminates in a loop. The upper hinged double-sided plate would have been attached to leather or some other material by means of one or two rivets.

There is one particular chape that does not look as though it was designed for such use. This is an example that dates to the Tudor period. This chape is thin and long – 4.5 inches in length by 0.5 inches in width. Another smaller example I have seen measures 3 inches in length by 0.25 inches in width. These were made from one piece of metal folded into a U-shape with an added thick terminal on the larger example.

The design on the smaller one is zigzag surrounded by a wriggle-work border; the larger example has a vine design that was favoured during this period and can be found on various belt fitments.

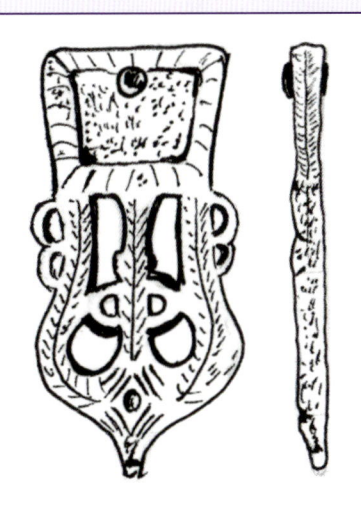

Fig.37. Medieval copper-alloy chape, amphora shaped openwork integral housing for the leather strap, one rivet.

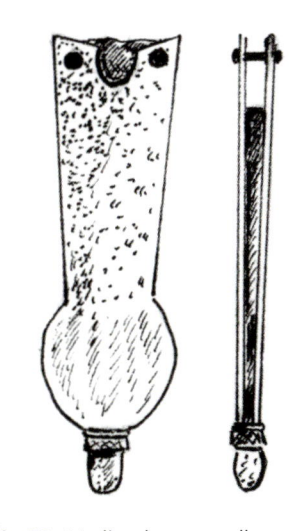

Fig.38. Medieval copper-alloy chape, crescent top oblong body lower being bulbous terminating in a acorn, two rivets.

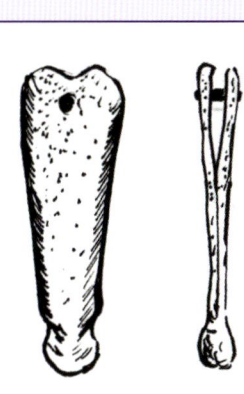

Fig.39. Medieval copper-alloy chape terminating in a knop, plain body, one rivet.

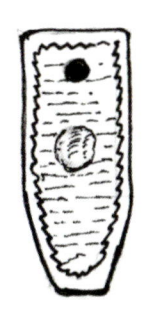

Fig.40. Medieval copper-alloy chape, folded sheet, central boss in a lined background surrounded by a wriggle work border, one rivet.

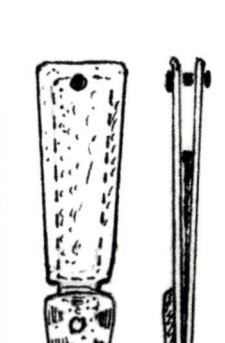

Fig.41. Medieval copper-alloy chape terminating in a raised square with radiate lines, plain body, one rivet.

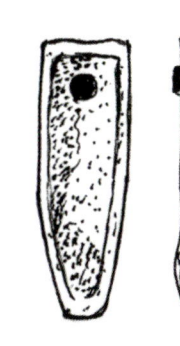

Fig.42. Medieval copper-alloy chape, folded sheet, plain body, lined border, one rivet.

Belt Chapes – Medieval

Fig.43. Medieval copper-alloy chape, openwork depicting the Golden Fleece, one rivet.

Fig.44. Medieval copper-alloy chape, folded sheet, black letter "R" above a roped line within an arch, two rivet holes.

Fig.45. Medieval copper-alloy chape, folded sheet, black letter within a lined border, two rivet holes.

Fig.46. Medieval copper-alloy chape, double line design within a lined border, two rivets.

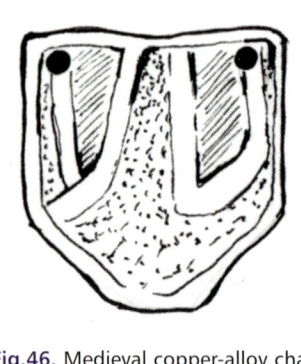

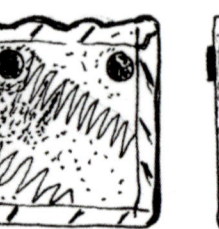

Fig.47. Medieval copper-alloy chape, wriggle work within a lined border, two rivets.

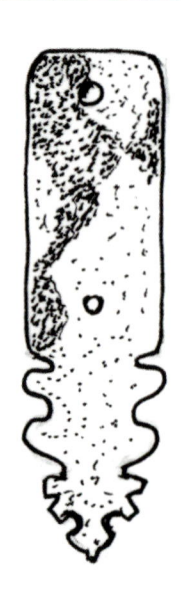

Fig.48. Medieval copper-alloy chape, plain body, scalloped lower half, two rivets.

Belt Chapes – Medieval

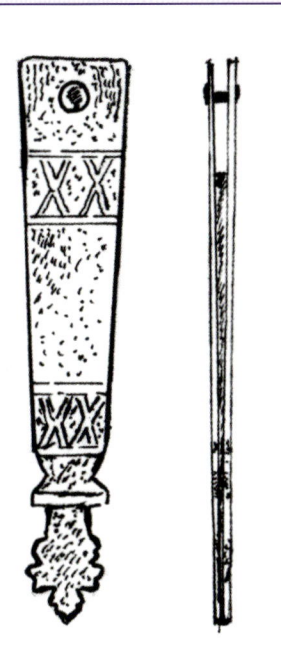

Fig.49. Medieval copper-alloy chape, two "Xs" within a lined border at the top, same design repeated on lower part terminating in a leaf-shaped knop, one rivet.

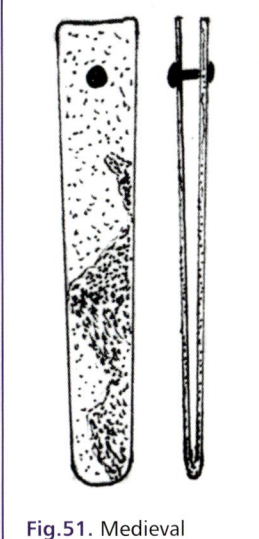

Fig.51. Medieval copper-alloy chape, plain body, one rivet.

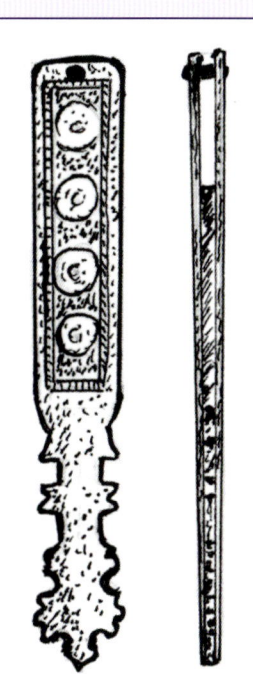

Fig.52. Medieval copper-alloy chape, top half four rings and dots surrounded by a beaded border, lower half multi cut out design border, one rivet.

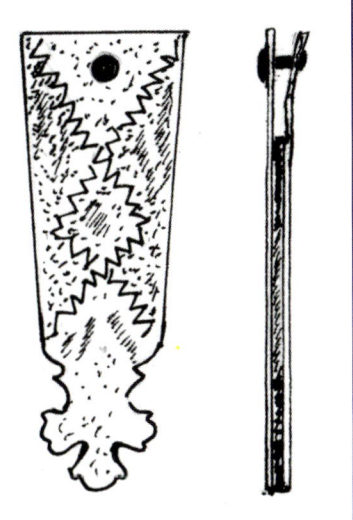

Fig.50. Medieval copper-alloy chape, wriggle work design terminating in a trefoil shape, one rivet.

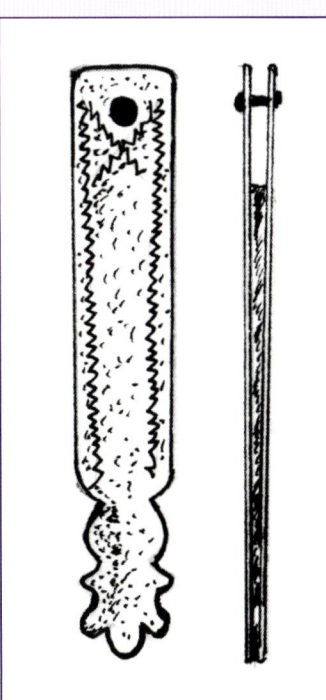

Fig.53. Medieval copper-alloy chape wriggle work "X" within a border terminating in a bulbous end with cinquefoil design, one rivet hole.

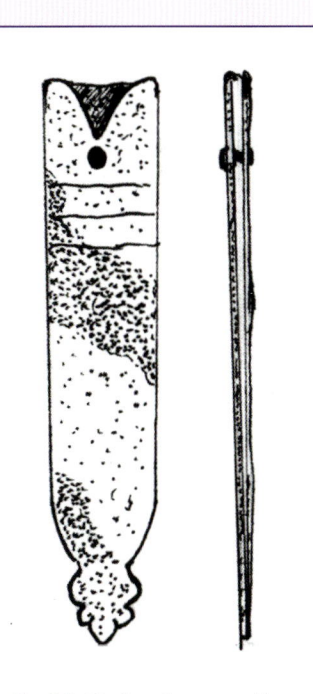

Fig.54. Medieval copper-alloy chape, "V" shaped top, three lines terminating in a cinquefoil design, one rivet.

Belt Chapes – Medieval

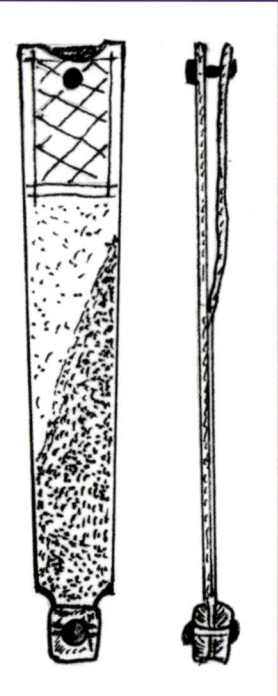

Fig.55. Medieval copper-alloy chape, cross hatch within a lined border terminating in a separate knop attached with a small band, one rivet.

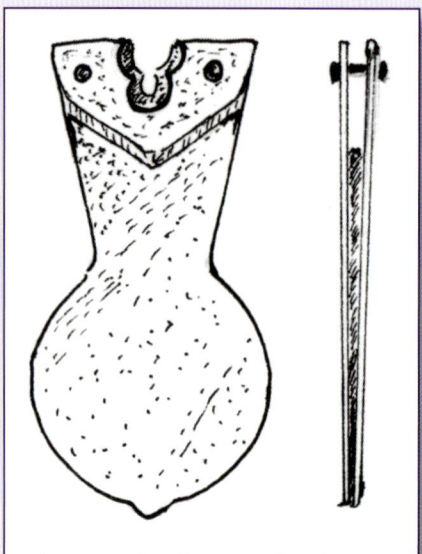

Fig.57. Medieval copper-alloy chape, trefoil top, single chevron design, bulbous body, two rivets.

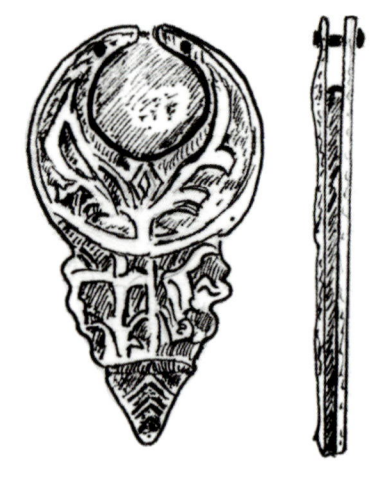

Fig.58. Medieval copper-alloy chape, almost an "O" shaped top, the whole having a foliate design, possibly a vine terminating in point with chevrons, two rivets.

Fig.59. Medieval iron chape, plain body terminating in a diamond-shaped knop below a band, three rivet holes.

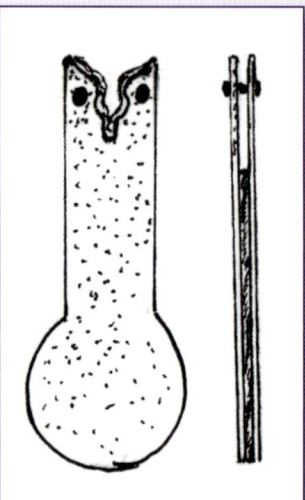

Fig.56. Medieval copper-alloy chape, curved open top, thin bulbous-shaped body, one rivet.

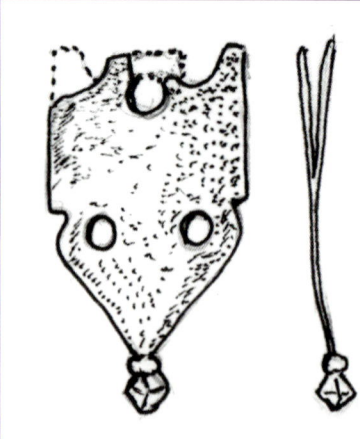

Fig.61. Medieval copper-alloy chape, shield shaped with turret top, plain body terminating in suspension loop with ring, one rivet.

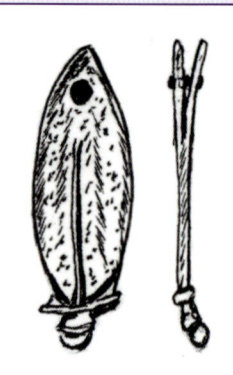

Fig.60. Medieval copper-alloy chape (these are often thought to be Saxon), oval body, vertical line within a border terminating in a horizontal band with a double knop, one rivet.

Belt Chapes – Medieval

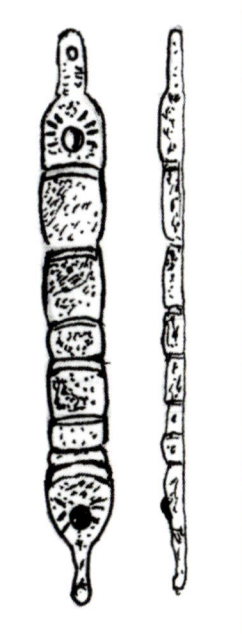

Fig.62. Medieval copper-alloy chape, seven segments divided by bands terminating in a thin knop, two rivet holes.

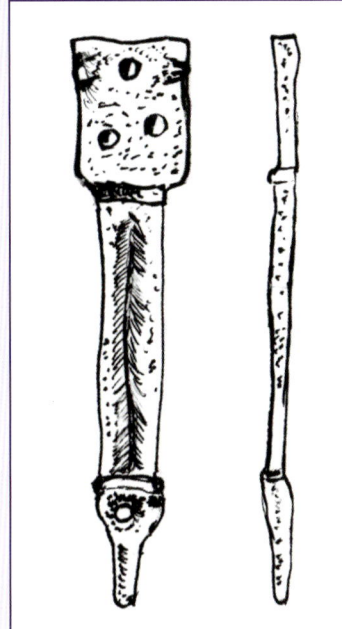

Fig.63. Medieval copper chape, square top, thin body with herring bone design terminating in a band with thin knop with hole, three rivet holes.

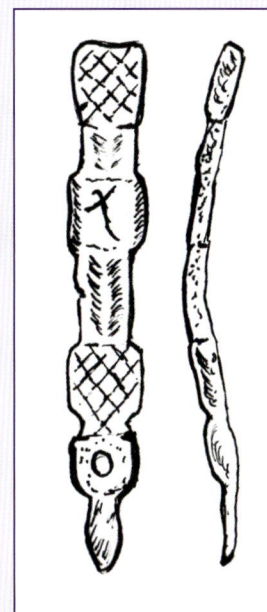

Fig.64. Medieval copper-alloy chape, three square segments, top and lower having cross hatching; centre an "X", terminates in knop with a hole, one rivet.

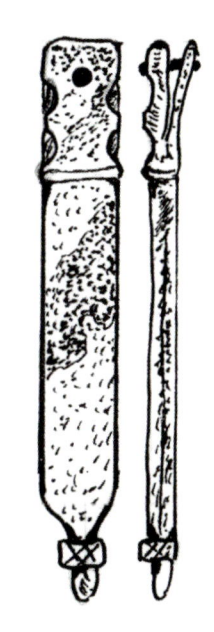

Fig.65. Medieval copper-alloy chape, top having curved sides above a single band collar, cross hatching above a knop, one rivet.

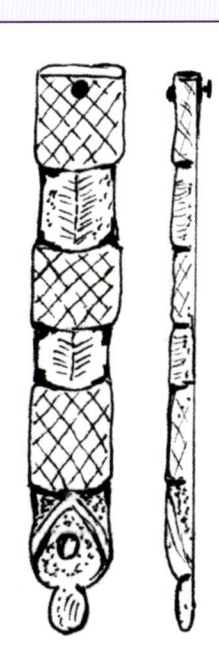

Fig.66. Medieval copper-alloy chape, three square segments each having cross hatching, terminating in a lined design with central hole and knop below, one rivet.

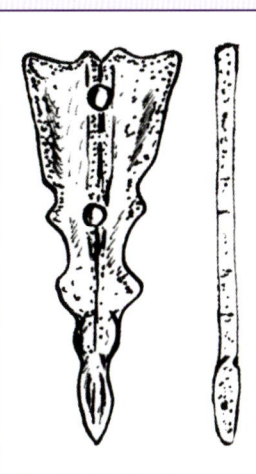

Fig.67. Medieval copper-alloy chape, "V" shaped body, scalloped sides terminating in a pointed knop, two rivet holes.

95

Belt Chapes – Medieval

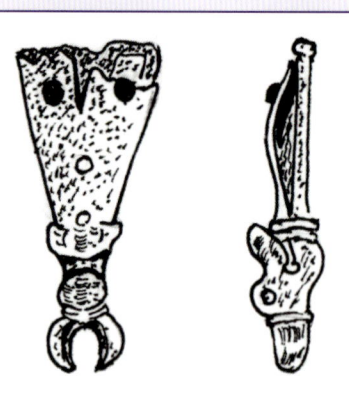

Fig.68. Medieval copper-alloy chape, "V" shaped body with two dots terminating in a beast's head, two rivets.

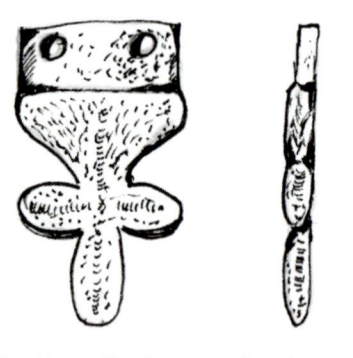

Fig.69. Medieval copper-alloy chape, plain trefoil body, two rivet holes.

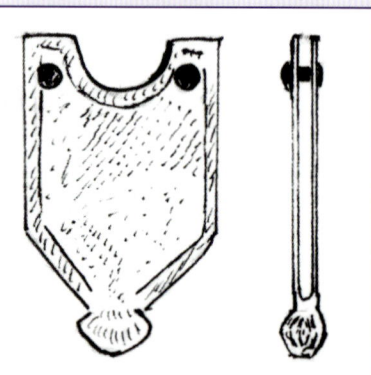

Fig.70. Medieval copper-alloy chape, shield shape with crescent cut out at top, lined border, terminates in knop, two rivets.

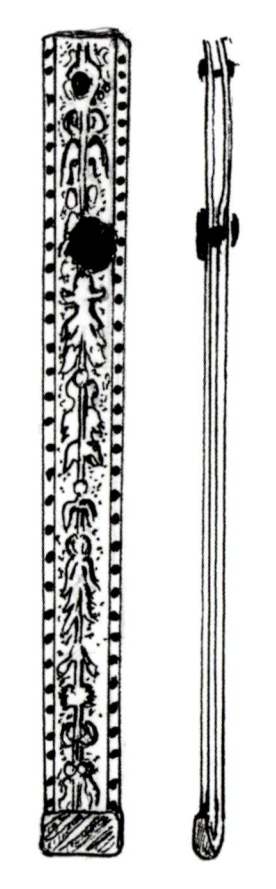

Fig.71. Tudor copper-alloy chape, vine design surrounded by a lined beaded border, terminates in a square knop, two rivets. (The lower part appears to be a replacement at a later date).

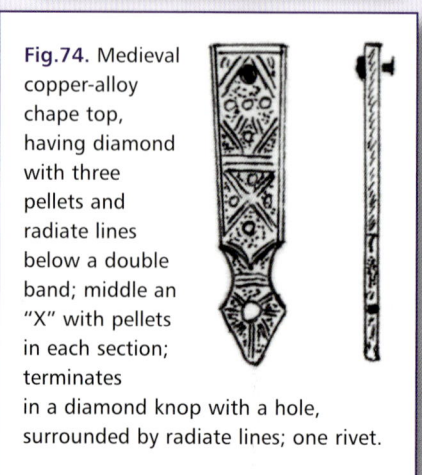

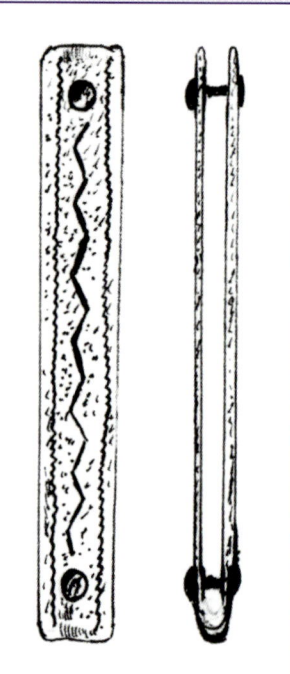

Fig.72. Tudor copper-alloy chape, zigzag design surrounded by a border, terminates in a knop, two rivet holes.

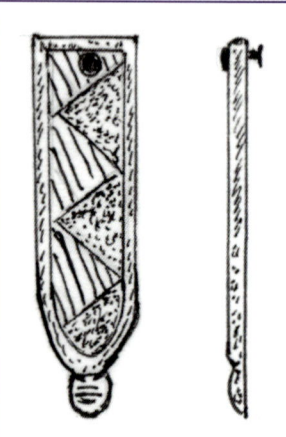

Fig.73. Medieval copper-alloy chape, zigzag design with one side having inner lines surrounded by a lined border, terminates, in a knop one rivet.

Fig.74. Medieval copper-alloy chape top, having diamond with three pellets and radiate lines below a double band; middle an "X" with pellets in each section; terminates in a diamond knop with a hole, surrounded by radiate lines; one rivet.

Belt Chapes – Medieval

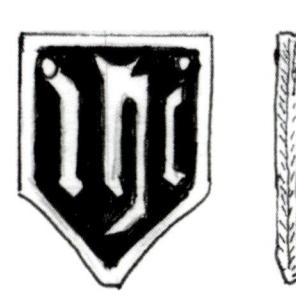

Fig.75. Medieval copper-alloy chape, shield shaped with the initials "IHC" (the first three letters of the Greek form of the name Jesus), two rivet holes.

Fig.76. Medieval copper-alloy chape, cross within a lined border, terminates in a scrolled trefoil knop, two rivets.

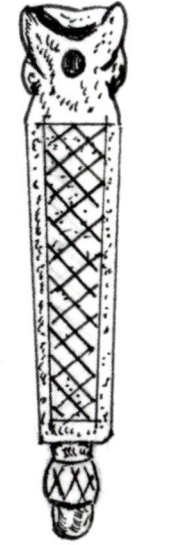

Fig.77. Medieval copper-alloy chape, crescent-shaped top with scrolled sides; crossed hatched centre within a lined border, terminates in an acorn knop, one rivet.

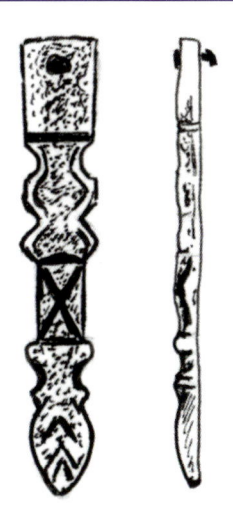

Fig.78. Medieval copper-alloy chape, square top, scrolled body divided by a central column with an "X", terminates in a knop with chevron design, two rivets.

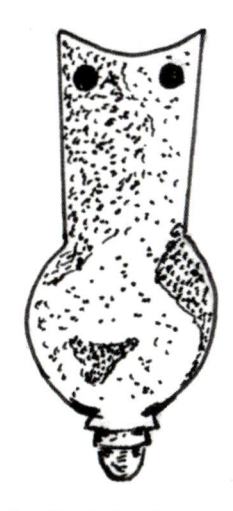

Fig.79. Medieval copper-alloy chape, plain bulbous body terminating in a knop, two rivets.

Fig.80. Medieval copper-alloy chape, worn lettering on the body (possibly "IHC") within a lined border, terminates in a leaf shape with radiate lines, two rivets.

Fig.81. Medieval copper-alloy chape, diamond design in centre with small pellet at top. This is flanked by four outer lines surrounded by a lined border; all terminating in a oval with central hole and radiate lines; one rivet.

97

Belt Chapes – Medieval

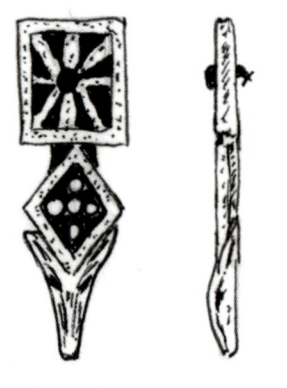

Fig.82. Medieval copper-alloy chape, square top central pellet with radiate lines; below this a diamond shape with five pellets in the form of a cross with lined border; all terminating in a beast's head (possibly a wolf); one rivet.

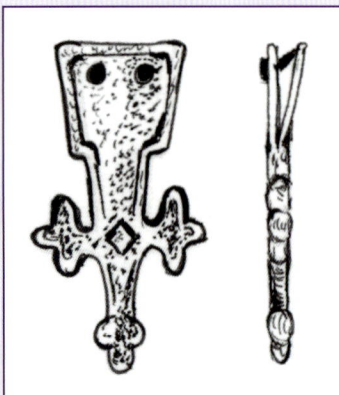

Fig.83. Medieval copper-alloy chape, square top, the whole terminating in a trefoil with protruding trefoil knops, lined border, two rivets.

Pic.1. Tudor copper-alloy chape, vine design surrounded by a lined beaded border, terminates in a square knop, two rivets. The lower part appears to be a replacement at a later date.

Fig.84. Medieval copper-alloy chape, shield-shaped, four-lined diamond design in centre, surrounded by a lined border, all terminating in two circles and a pointed knop, two rivets.

Fig.85. Medieval copper-alloy chape, shield-shaped, cross hatching within two horizontal lines, two rivets.

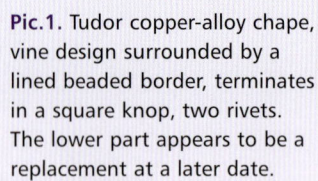

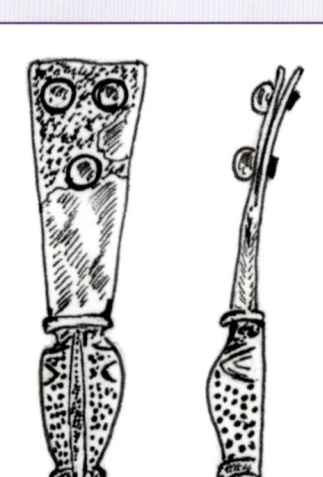

Fig.86. Medieval copper-alloy chape, top half plain but terminating in a wolf's head covered with small indentations and having raised ears; the snout has a small suspension hole. Three raised rivets, also of copper-alloy; the whole of this chape still retains much of its gilding.

Belt Chapes – Medieval

Pic.3. Left: Medieval cooper-alloy chape, folded sheet, black lettering within lined border, two rivet holes. Right: Medieval copper-alloy chape decorated with interlocking crudely-shaped chevrons, two rivets.

Pic.4. Medieval copper-alloy chape, "V" shaped body, scalloped sides terminating in a pointed knop, two rivet holes.

Pic.5. Medieval copper-alloy chape, square top, thin body with herring-bone design, terminates in a band with thin knop, three rivet holes.

Pic.2. Various copper-alloy chape spacers.

Pic.6. Medieval copper-alloy chape, three square segments, top and lower having cross hatching, the centre having an "X", the whole terminating in a knop, one rivet hole.

Pic.7. Medieval copper-alloy chape, "V" shaped body with two dots, terminates in beast's head, two rivets.

Pic.8. Medieval copper-alloy chape, folded sheet with turned ends, geometric design, two rivet holes.

Belt Chapes – Medieval

Pic.9. Medieval copper-alloy chape, folded sheet, black letter "R" above a roped line within an arch, two rivet holes.

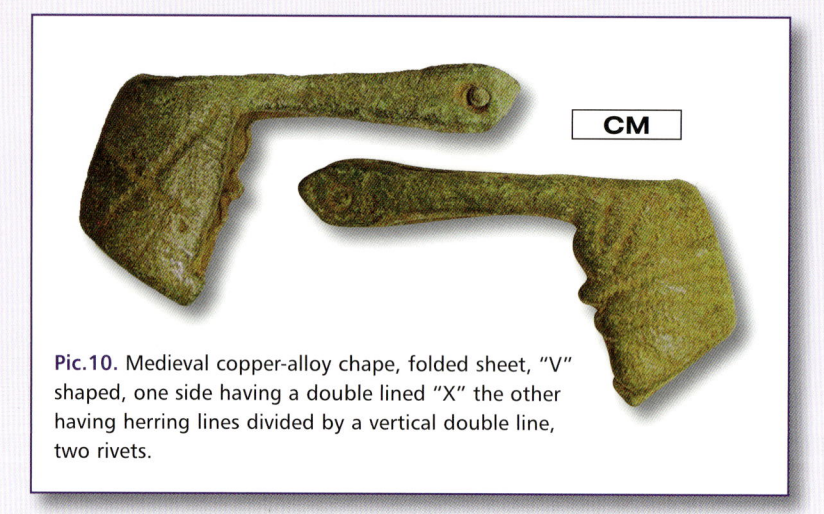

Pic.10. Medieval copper-alloy chape, folded sheet, "V" shaped, one side having a double lined "X" the other having herring lines divided by a vertical double line, two rivets.

Pic.11. Medieval copper-alloy chape (these are often incorrectly thought to be Saxon), oval body having a vertical line within a border; terminates in a horizontal band within a double knop, one rivet.

Pic.12. Medieval copper-alloy chape, shield-shaped with lined border, one rivet.

Pic.13. Medieval copper-alloy chape, plain body terminating in a knop, one rivet.

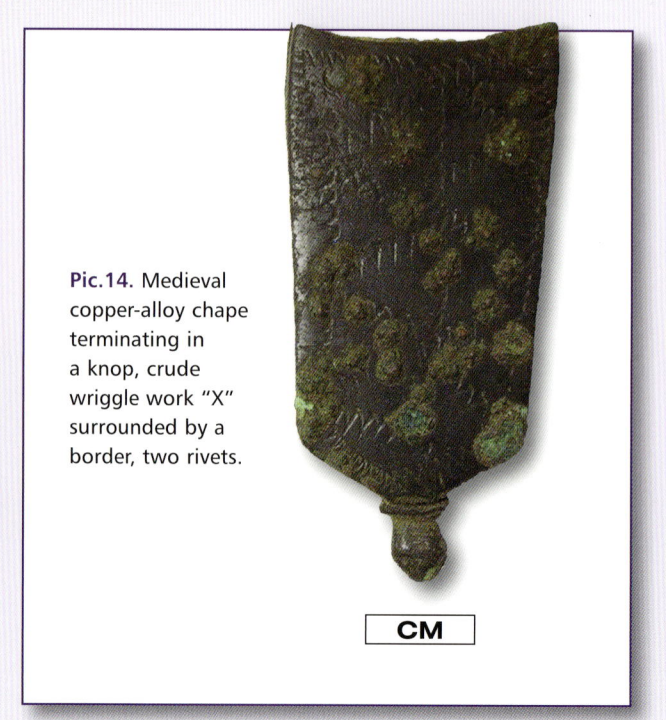

Pic.14. Medieval copper-alloy chape terminating in a knop, crude wriggle work "X" surrounded by a border, two rivets.

Belt Chapes – Medieval

Pic.15. Medieval copper-alloy chape terminating in a trefoil, crossed wriggle above a double line, single rivet.

Pic.16. Medieval copper-alloy chape, plain body, one rivet.

Pic.17. Medieval copper-alloy chape, plain body, crescent-shaped top, two rivets.

Pic.18. Medieval copper-alloy chape, plain body, one rivet.

Pic.19. Medieval copper-alloy chape terminating in a knop, wriggle work oblong scroll surrounded by a wriggle work border, one rivet.

Pic.20. Medieval copper-alloy chape, crescent-shaped top, plain body, two rivet holes.

Pic.21. Medieval copper-alloy chape, plain body, single rivet.

Pic.22. Medieval copper-alloy chape, plain body, single rivet.

Belt Chapes – Medieval

Pic.23. Medieval copper-alloy chape, crescent-shaped top, plain body, two rivets.

Pic.24. Top part of a medieval bronze chape, black lettering, two rivets.

Pic.25. Medieval copper-alloy chape, wriggle work "X", terminating in a knop, two rivets.

Pic.26. Medieval copper-alloy hinged two piece chape, swivel type, radiate lines around edge, two rivets.

Pic.27. Medieval copper-alloy chape, worn lettering on the body, possibly "IHC", within a lined border, terminates in a leaf shape with radiate lines, two rivets.

Pic.28. Medieval copper-alloy chape. The top half is plain, the lower terminating in the head of a wolf covered by small indentations and having raised ears; the snout has a small suspension loop. The three raised rivets are bulbous shaped. The whole of the chape still retains much of its gilding.

102

Medieval Gilded Studs

Various sizes of iron and copper-alloy studs (as opposed to small tacks) have been recovered both during excavations, and as single finds on fields. This is usually with the aid of a metal detector, but on the odd occasion "eyes only".

They are normally recovered where a dwelling once stood, or near holy places such as churches although sometimes as isolated finds. The former is understandable for where people lived or worked then losses occur. The same applies to the workshops where the studs were originally made.

As to holy places finds of studs could be attributed to losses by those who where associated with them. Also, with the Dissolution of the Monasteries, many such holy places were despoiled by Henry VIII's soldiers. Non-precious items would have been discarded and thrown into the surrounding fields, some being smashed to pieces and others burnt. All that now remains would be the metal that was at one time attached to them.

Another possibility is robbery, for example a cask that was taken and broken into away from the scene of the crime.

However, on the rare occasion some of these studs still show the remains of their original gilding, and on a few examples this gilding is almost completely preserved. Many of these studs have a domed-shaped head, prominent or just slightly domed. The shafts also vary both in thickness and length.

Judging from the examples that I have been able to study, the whole stud was gilded as opposed to just the head (which was the only part to be seen).

Many of the early medieval examples were plain, with only a few of the later medieval studs carrying some form of minimal design.

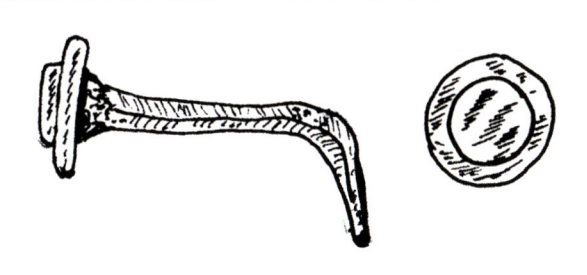

Figs.1a & b. Copper-alloy stud with raised flat top. The shaft is square in section and straight until the bend, which indicates the thickness of the wood or leather that it was once attached to. The whole of this stud was at one time gilded, and the top section still retains some of this.

Figs.2a & b. Show a stud that would have been attached to either wood or leather. Many studs of this type were gilded and quite heavy.

Figs.4a & b. Small copper-alloy stud, the top being domed shaped. Much of its original gilding remains.

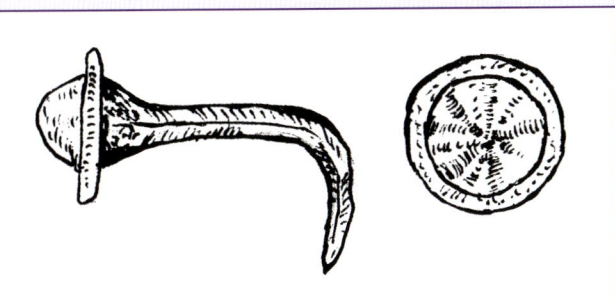

Figs.3a & b. This copper-alloy stud is similar to Fig.1., except the top is domed shaped. The dome still retains all of its gilding.

Figs.5a & b. Small copper-alloy stud the top being domed shaped, and its shaft square in section. Much of its gilding remains.

Medieval Gilded Studs

As to what these specific studs were attached to – in view of the cost and labour involved in completely gilding them – it is now only possible to surmise. The most likely deduction is probably ornate reliquary containers used for housing various holy relics.

Many medieval monasteries would have acquired and displayed such containers and what they thought to be the holy relics they contained. Not only would the monks feel that the relics watched over them, but also they would have created a great income for the monasteries. If famous enough, pilgrims would pay to come and see or touch the relics concerned with the hope that God would look favourably on them, or that the cure for some ailment might be forthcoming.

Besides their religious connotations, it is also possible that gilded studs were used as decoration on furniture intended for the nobility or those wealthy enough to have afforded such ostentation.

Some of these studs have small indentations on the head – seemingly as a result of where they have been hammered into wood. However, if an iron hammer had been used then it is likely some of the gilding would have been damaged. It is therefore likely that they would have been covered with leather to protect them during this process.

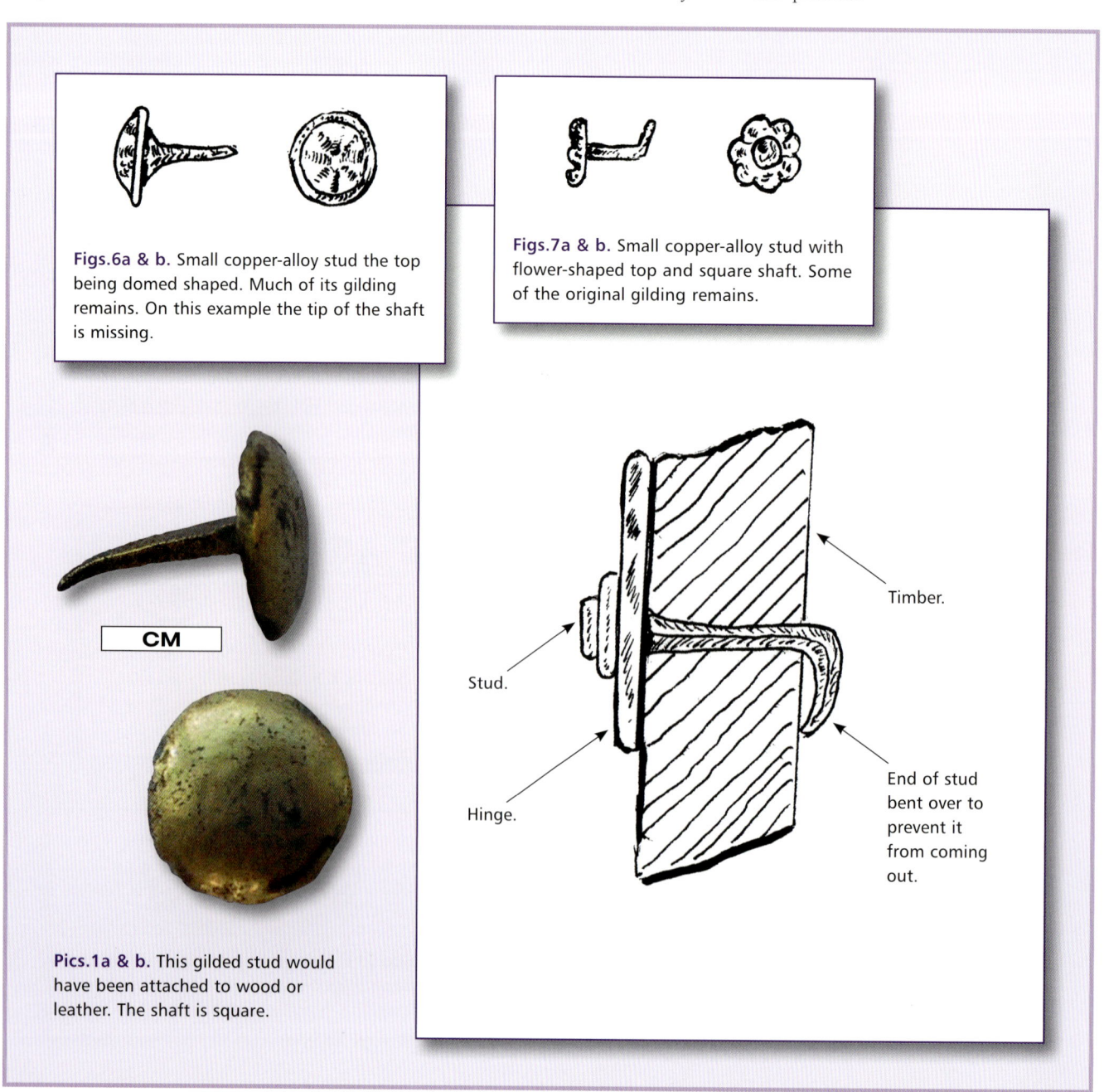

Figs.6a & b. Small copper-alloy stud the top being domed shaped. Much of its gilding remains. On this example the tip of the shaft is missing.

Figs.7a & b. Small copper-alloy stud with flower-shaped top and square shaft. Some of the original gilding remains.

Pics.1a & b. This gilded stud would have been attached to wood or leather. The shaft is square.

Stud.

Hinge.

Timber.

End of stud bent over to prevent it from coming out.

104

Non-Heraldic Bells

The small, round copper-alloy button-shaped bells that sometimes turn up as detector finds have often been wrongly described as intended for use on shoes, jackets or hats – all such purposes being related to the adornment of human beings.

Nothing could be further from the truth, for these artefacts are strictly harness decorations. They come under the banner of "non-heraldic" pendants as they bear no coat of arms. The bell would have attached to a mount, similar to those used on the normal heraldic and non-heraldic pendants. Such mounts might be of single circular form, having either a single or double rivet for securing it to leather, or a T-shaped bar with studs at either end. Both would have had an open central hinge. Another known example of such mounts is of ball form with a lower tubular securing projection. Four stems project from the ball, each being "L" or dog leg in shape and terminating in an open central hinge.

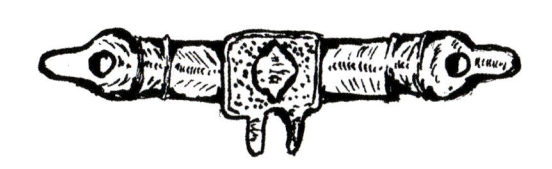

Fig.1. Pendant/bell mount, each end having a hole for fastening with a rivet. The centre has an open hinge to which the pendant or bell would have been attached.

Fig.2. Pendant/bell mount, circular in shape with a stud for fastening to leather. The open hinge would have been used for securing the pendant or bell.

Fig.3. Pendant/bell mount, oblong in shape with two studs for fastening to leather. The open hinge would have been used for securing the pendant or bell.

Fig.4. Harness fitting in the shape of a bell with a lower tubular protrusion. The four independent arms are "L" or dog leg in shape, each having a hole to which the bell or pendant would have been attached.

Fig.5a & b. Bell with long shank but small body. Vestiges of original gilding remain.

Non-Heraldic Bells

Fig.6a & b. Bell with medium length shank but large body; signs of original gilding.

Fig.7a & b. Bell with comparatively large body and long shank.

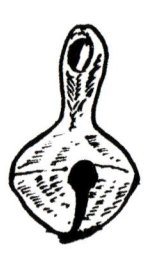

Fig.8a & b. Bell with large body and shank of medium width.

Fig.9a & b. Damaged bell with large body and long shank.

Fig.10a & b. Damaged bell with large body and round shank; signs of black pitch.

Fig.11a & b. Although the size of a normal non-heraldic bell, this example is in fact a small folly bell used by people for adorning their clothing; it dates to the 15th century.

Fig.12a & b. Although similar to a non-heraldic bell, this example is a hawking bell.

Non-Heraldic Bells

The mounts could be either decorated or plain, though to date it would appear that the latter is more common.

The diameter of these bells ranges from half an inch to three-quarters of an inch. The length of the shanks also varies greatly from a quarter of an inch to an inch, but all terminate in a small loop.

The small pea inside the bell was made from iron, or the same metal as the bell itself.

Some of these bells still retain traces of gilding, although tinning was probably also common. Black pitch has been found on a few examples, but this treatment seems to have been fairly rare. The gilding or other forms of coating would also have been applied to the mount, regardless of style.

Small bells of this type should not be confused with hawking bells, folly bells, or those used for children's rattles. The main distinguishing feature is that pendant bells were usually more crudely produced.

The construction of these bells differs greatly from other types. They are made from a single thin sheet of copper, each section being cut and shaped

Pic.1. When parted company from their metal strap these hawking bells can often be mistaken for non-heraldic bells. CM

Pic.2. Pendant/bell mount. The two holes at each end of the bar were used to secure the mount to leather. CM

Pic.3. Bell with large body, shank of medium width. CM

Pic.4. Pendant/bell mount, ornate circular body, stud on back. The open hinge was used for securing either a bell or pendant. CM

Pic.5. Bell with damaged body and long shank. CM

107

Non-Heraldic Bells

into an envelope design. The four sections are then folded over to create a round shape, the pea being added at the last fold; the loop is attached last.

Another type of medieval harness pendant bell is pear or tear-drop shaped. These would have been used in the same manner as the round examples, possibly with the exception of the four stem types. The construction of the pear-shaped bell differs slightly, in that the whole artefact was made from a single sheet of copper. The shape was created by folding the metal slightly so that it overlapped along the entire length – wide at the base but narrowing at the top. The loop was then formed by hammering the top flat, and making a suspension hole. Both types of such pendant bells date 13th-14th century. As with the non-heraldic types they were much favoured in their time, not only for the comfort of the pleasant sound they made but also because the bell was regarded as a method for keeping away the evil eye.

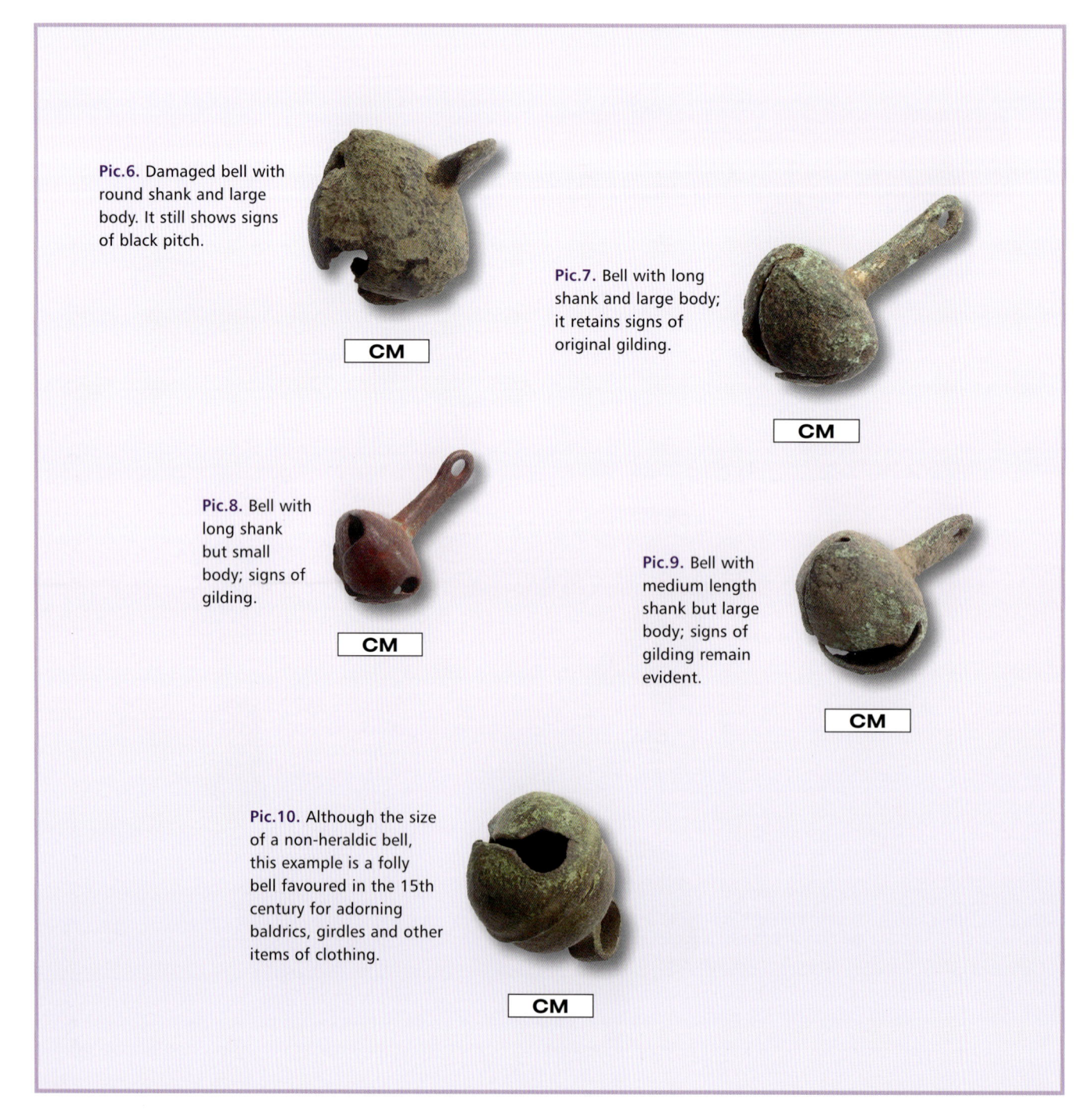

Pic.6. Damaged bell with round shank and large body. It still shows signs of black pitch.

Pic.7. Bell with long shank and large body; it retains signs of original gilding.

Pic.8. Bell with long shank but small body; signs of gilding.

Pic.9. Bell with medium length shank but large body; signs of gilding remain evident.

Pic.10. Although the size of a non-heraldic bell, this example is a folly bell favoured in the 15th century for adorning baldrics, girdles and other items of clothing.

Lead Weights

Lead has been used in the making of trade weights for many centuries. In this country, in fact, right back to the Roman period. At this time weights were widely used throughout the land, and would have been accepted to be of the correct value.

The intended use of some of these weights can be determined (i.e. conical with a top hole or moulded-in loop to operate with a steelyard); others were round and flat for use with normal balance scales.

Some known examples have moulded or engraved lines to indicate their weight (e.g. one line for 1oz, two lines for 2oz and so forth).

When the Romans left these shores the way of life they had introduced did not suddenly end, neither did their way of trading. Weights manufactured just after the Romans had left would have been made in the same style as their predecessors.

Not a great deal is known about the weights used in the Anglo-Saxon and Viking periods, but they did exist and there are known examples. Some were simply adapted from worn Roman coins, with marks made on them to indicate value (e.g. ring and dots or annulets). Others were more sophisticated and purpose made from bronze or lead, or a combination of both metals.

Some lead weights from this period are also known with strange undecipherable inscriptions on one side. Others are flat and round without any inscriptions or attempts at design.

Improvements came in the Norman period, and there are some distinctive

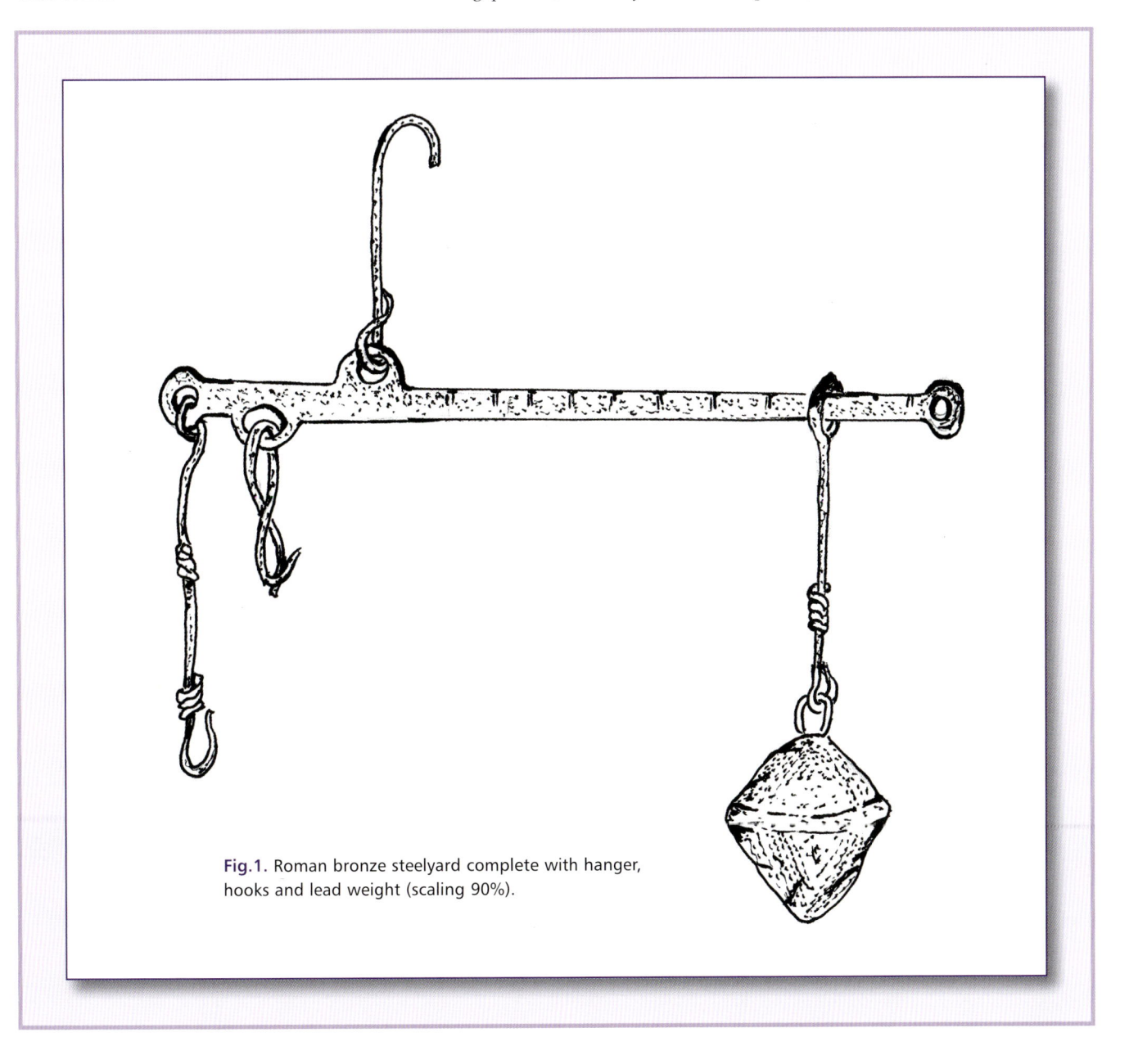

Fig.1. Roman bronze steelyard complete with hanger, hooks and lead weight (scaling 90%).

Lead Weights

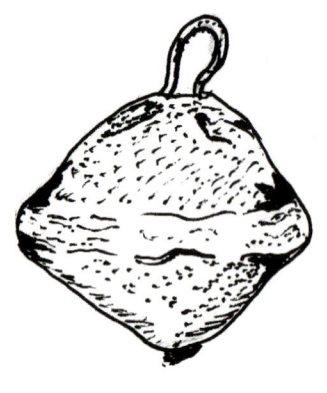

Fig.2. Roman lead steelyard weight with raised line around the perimeter. This example has a copper suspension loop.

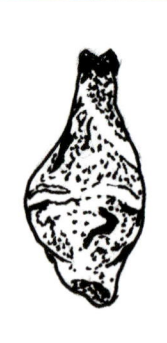

Fig.3. Roman lead steelyard weight with remains of an iron loop.

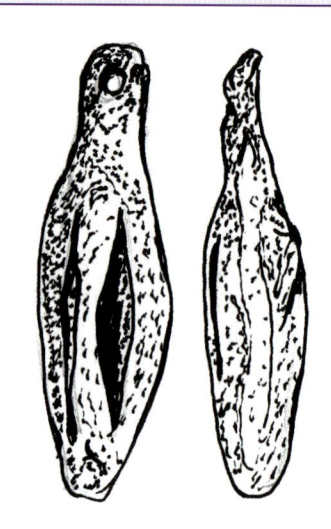

Fig.4. Unusual Roman lead steelyard weight, each side having two oblong concave lines; holed at top.

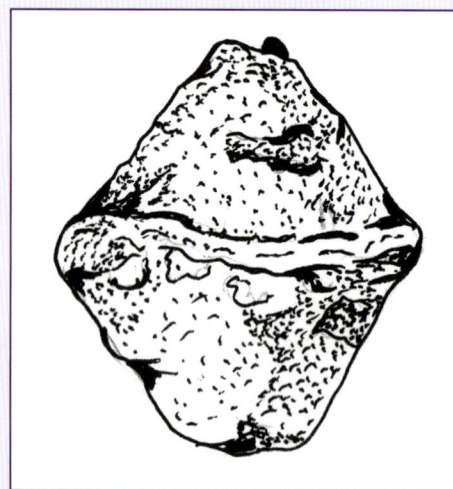

Fig.5. Large lead steelyard weight, raised line around perimeter, remains of iron loop.

Fig.7. Medieval lead trade with merchants mark "M" or "W".

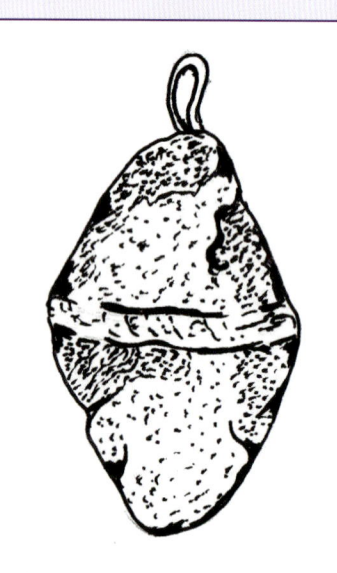

Fig.6. Roman lead steelyard weight raised line around perimeter. It has a copper suspension loop.

Lead Weights

types and designs. One group in particular had a floral pattern or the design of a floriated cross. Such examples were well executed, and the casting uniform. Some had an outer decorated border and look similar to a matrix seal in appearance.

Medieval and post medieval seals are hard to date unless there is a mark showing the reigning monarch or pertaining to a particular city or town. However, some early home made lead weights can be identified and dated if they carry a design. For example, a very large weight in my collection carries a lead long cross with a pellet in each corner. The person or workshop responsible for this weight has based the design on that of a long cross penny thereby making it appear official.

Many medieval home made round weights are very crude – some having a raised rim (on one side or both), with others being flat. Another type was the conical free-standing weight. These show a wide variation in diameter and height, and some have a hole at the top. The use of the purpose-made hole is open to supposition. Did a set of weights have a cord passed through them to make them easier to carry while travelling? Or were they part of some type of medieval steelyard? What can be said is that none of the examples I have seen carry any designs.

During the late 14th century and

Fig.8. Medieval lead trade weight showing fleur-de-lis with pellets below a crown within a lined border.

Fig.9. Medieval lead trade weight with the cross of St Andrew.

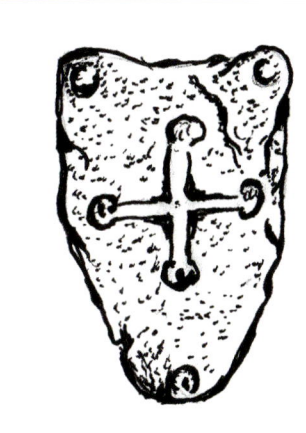

Fig.10. Medieval lead trade weight, cross with three pellets.

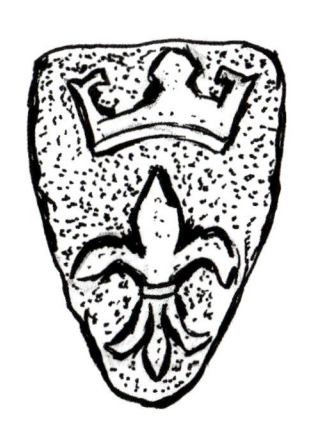

Fig.11. Medieval lead trade weight, fleurs-de-lis below a crown.

Fig.12. Medieval lead trade weight with what is possibly a lion rampant within a roped border.

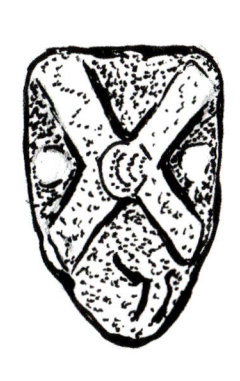

Fig.13 Medieval lead trade weight, cross of St Andrew with central pellet and pellets right and left.

Lead Weights

Fig.14. Medieval lead trade weight with merchant's mark within a lined border.

Fig.15 Medieval lead trade weight with fleur-de-lis.

Fig.16. Medieval lead trade weight with merchant's mark.

Fig.17. Medieval lead trade weight, three lions passant within a lined border; hole at top.

Fig.18. Medieval lead trade weight with merchant's mark within a lined border.

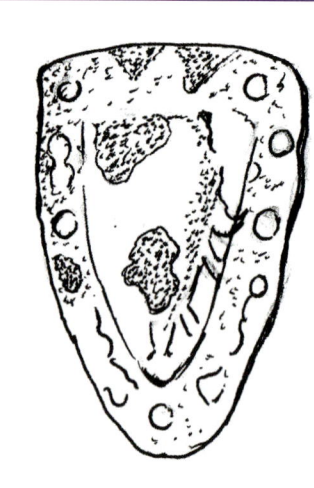

Fig.19. Medieval lead trade weight, worn raised shield design surrounded by pellets.

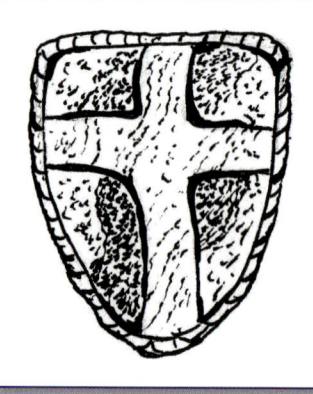

Fig.20. Medieval lead trade weight, raised cross surrounded by a roped border.

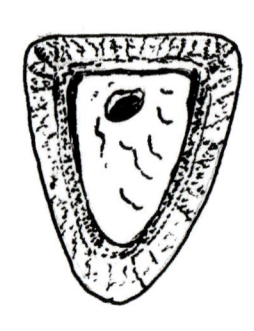

Fig.21. Medieval lead trade weight, raised plain shield with a rope edge; holed at top.

Lead Weights

into the early 15th century, the flat round lead weights had to conform to the avoirdupois system and were stamped with an incuse crown to show that they were of good weight. This mark may have been used by the Grocers Company, and possibly the first attempt to regulate the avoirdupois weight.

Although there were official bronze weights available that could – and should – have replaced the lead examples, the latter remained in use for many centuries…especially in rural areas. This can be verified by the amount excavated bearing official stamps such as the initial or initials of the reigning monarch or monarchs. Occasionally, there will be two or even three monarchs' initials on a lead weight. This was because when checked over a period of time should the weight still be good then the initials of the reigning monarch would be stamped on the lead weight to show such. If the weight was checked and found not to be good, the maker would be told to discard it. However, as so many are recovered without any official marks it is more than likely that the weight remained in use and officialdom disregarded.

Small flat lead weights ranging from 12mm to 16mm square have been recovered, in addition to round flat examples bearing the symbols that are normally associated with the weighing

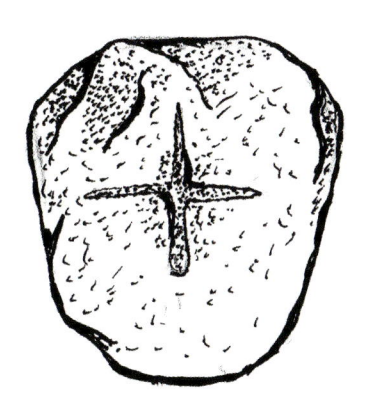

Figs.22a & b. Medieval lead trade weight showing a crude crown; the reverse carries a small cross.

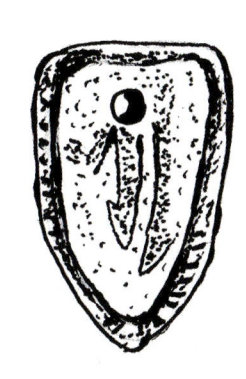

Fig.23. Medieval lead trade weight, worn design surrounded by a radiate border; holed at top.

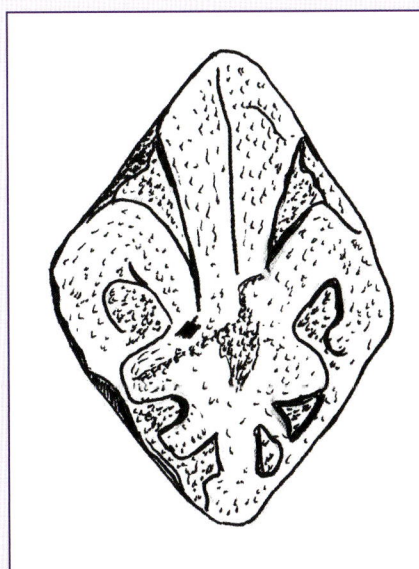

Fig.24. Medieval lead trade weight, the whole showing a fleur-de-lis.

Fig.25. Medieval lead trade weight, raised cross similar to London mark though without the dagger.

Fig.26 Medieval lead trade weight, cross of St Andrew within a lined border.

113

Lead Weights

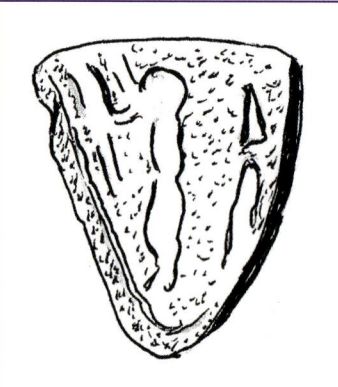

Fig.27. Medieval lead trade weight, possibly a lion rampant within a lined border.

Fig.28. Medieval lead trade weight showing fleur-de-lis.

Fig.29. Medieval trade weight the whole being of a fleur-de-lis shape.

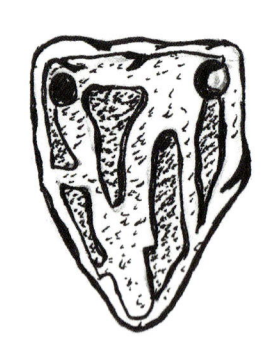

Fig.30. Medieval lead trade weight with unclear design (possibly a merchant's mark); two holes at top.

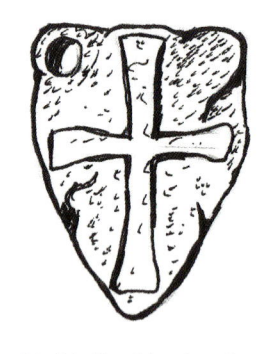

Fig.31. Medieval lead trade weight showing a cross; hole at top.

Fig.32. Medieval lead trade weight showing three unidentified shapes within a lined border surrounded by pellets.

Fig.35. Medieval lead trade weight, six-pointed star within a sexfoil. Unlikely to have been a coin weight due to its heaviness.

Fig.34. Medieval lead trade weight, fleur-de-lis within a double lined diamond shape. Although of coin weight size this is unlikely to have been the case due to its heaviness.

Fig.33. Medieval lead weight of quatrefoil shape, central pellet with four radiate double lines, three of the quarters having a single pellet.

114

Lead Weights

of coins. It is very doubtful that these would have been anywhere near the correct weight of the official examples; like many home made weights they were produced to deceive.

Should they have been questioned, then the user only had to point to the "official" markings hoping the matter would not be pursued any further.

Shield-shaped lead weights are rare and not many have survived. They give the appearance of official weights purely from their shape and design. It is possible that they were produced in the latter part of the 14th century following the conversion to the trade or avoirdupois pound. Some have an emblem such as a fleur-de-lis (with or without a crown above), a raised plain shield, or even an heraldic design normally found on horse pedants from that period such as three lions passant or lion rampant; others have English or Scottish crosses.

Lead is very soft and when used constantly then any raised design would wear away, sometimes leaving a blank surface.

The weight used for this type appears to have varied considerably, ranging from small examples (e.g. 5, 6, 7, 8, 9, 10, 12 or 14 ounces), to the larger types weighing between one and two pounds. This shows that these shield-shaped weights, along with other shapes, were often made by traders who were out to catch the public.

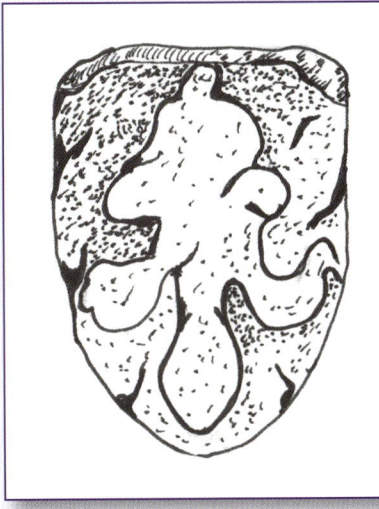

Fig.36. Medieval lead trade weight, very crude fleur-de-lis which is upside down.

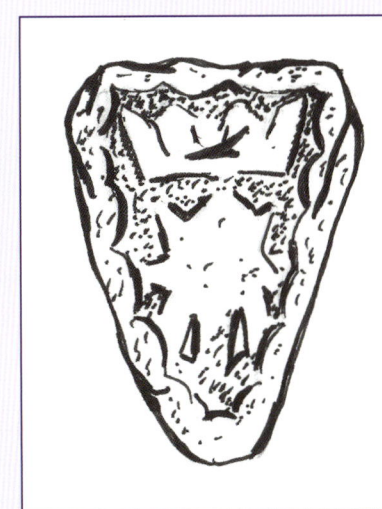

Fig.37. Medieval lead trade weight unrecognisable (possibly a merchant's mark) design below a crown.

Fig.38. Medieval lead trade weight unusual central design which appears to be a double lined over lapping circle with cross and scales. Lettering in the outer part most of which is illegible.

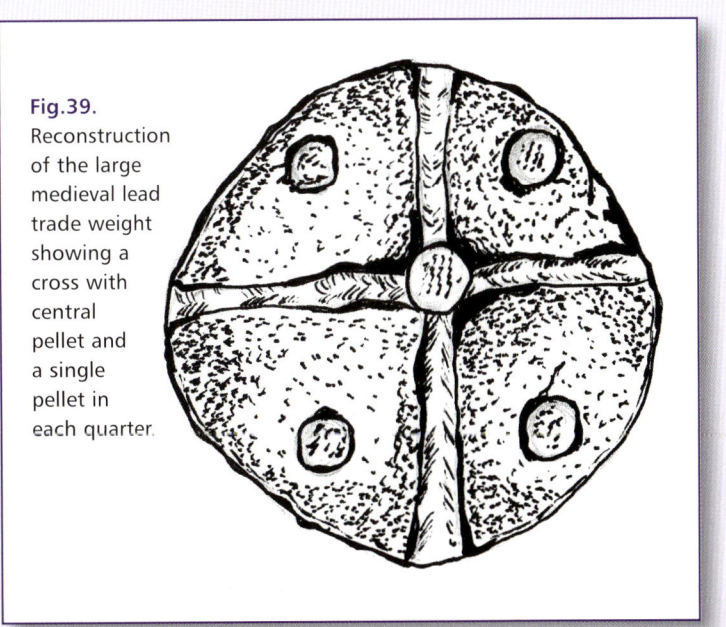

Fig.39. Reconstruction of the large medieval lead trade weight showing a cross with central pellet and a single pellet in each quarter.

Lead Weights

Fig.40. Medieval lead trade weight, four leaf design within a quatrefoil.

Fig.41. Medieval lead trade weight, cross with a pellet in each quarter.

Fig.42. Lead trade weight, post medieval, nine spoke wheel.

Fig.43. Medieval lead trade weight, four equal concave sections which create the effect of raised cross with border.

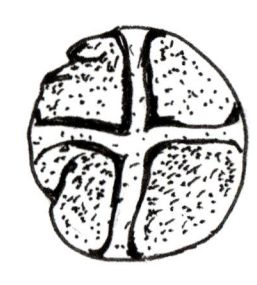

Fig.44. Medieval lead trade weight showing crude long cross.

Fig.45. Medieval lead trade weight, long cross with radiate lines in each quarter.

Fig.46. Medieval lead trade weight, cross made up of zigzag lines.

Fig.47. Medieval lead trade weight, circle of leaves.

Fig.48. Lead trade weight, post medieval, initials "CL" below "IV", raised rim.

Fig.49. Lead trade weight, six pointed sun burst, possibly 16th-17th century.

Lead Weights

Fig.50. Lead trade weight with design almost representing a seal matrix. The centre with possible letters and the edge with what appears to be an inscription.

Figs.51 & 52. Lead trade weights. One has three lines the other two. These appear to have been added after casting. Similar examples have been recovered on sites ranging from the Roman period through to medieval.

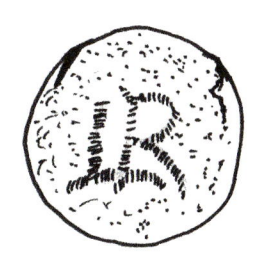

Fig.53 Lead trade weight with initials "IR" the letter "I" being a "J" scratched on after casting; post medieval.

Fig.54. Lead trade weight, post medieval, letters "HC" above two crescents.

Fig.55. Lead trade weight, post medieval, with three lined circles the outer having illegible lettering.

Fig.56. Medieval lead trade weight, centre pellet with radiate lines.

Fig.57. Lead trade lead weight, post medieval, radiate star surrounded by three lined circles.

Fig.58. Medieval lead trade weight, octofoil shaped, having a central pellet with radiate leaves.

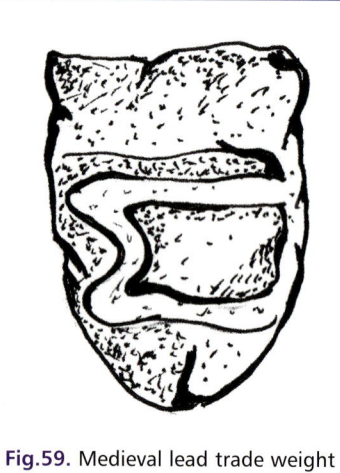

Lead Weights

Fig.59. Medieval lead trade weight with possible merchant's mark.

Fig.60. Various small medieval lead trade weights. These could have been used in conjunction with larger weights.

Fig.61. Medieval lead trade weight having the cross of St Andrew with two pellets in each quarter. This could have been used as a single weight or in conjunction with larger examples.

Fig.62. Medieval lead trade weight having five irregular petals within a lined border.

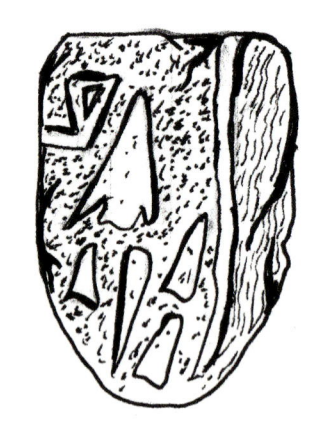

Fig.63. Medieval trade weight, irregular shapes with vertical double line and possible merchant's mark.

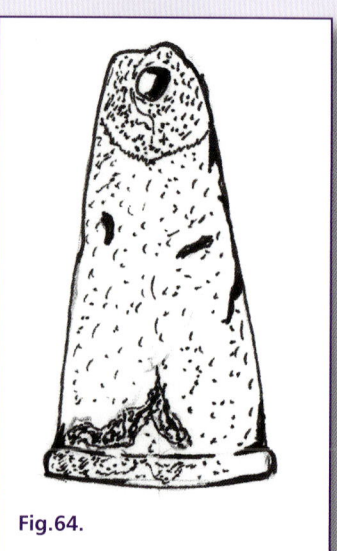

Fig.64.

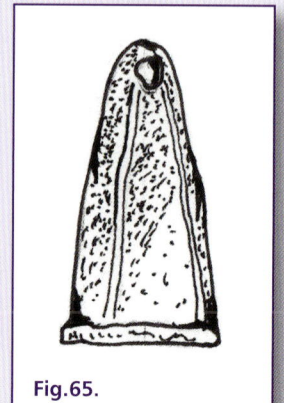

Fig.65.

Fig.66.

Figs.64-66. Three lead conical weights. These could have been used free standing or with a steelyard. Such examples were in use from the Roman period through to post medieval times.

Lead Weights

Fig.67. Lead trade weight of Elizabeth I, "EL" below a crown with the dagger to the right (London).

Fig.68. Lead trade weight of Charles II "CIIR" below a crown within a circle; to the right Norwich mark (Norfolk).

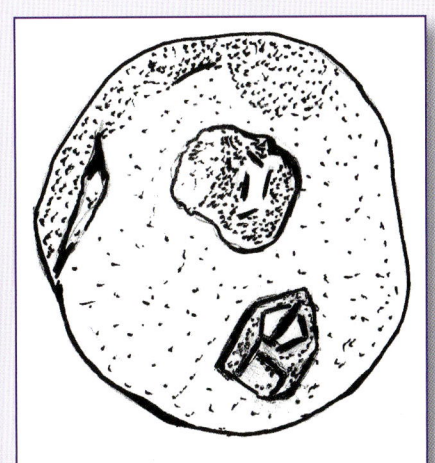

Fig.69. Lead trade weight of either Henry VII or VIII "h" below crown with possible plumber's mark above.

Fig.70. Lead trade weight of Elizabeth I "EL" within a dome-shaped raised border.

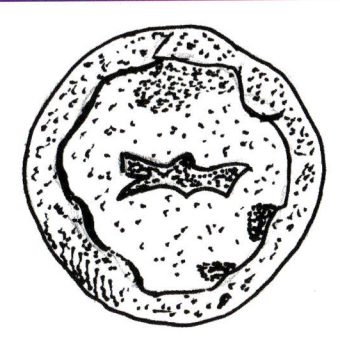

Figs.71a & b. Lead trade weight, post medieval, one side showing what could be some bird in flight, the other side an irregular shape. The edge shows an attempt has been made to produce a milled effect similar to that of coin.

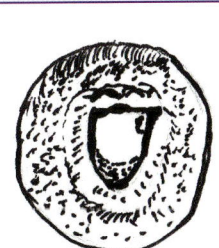

Figs.72a & b. Lead trade weight, post medieval, one side having a shield shape surrounded by a faint outer line; the other side shows a distorted shape.

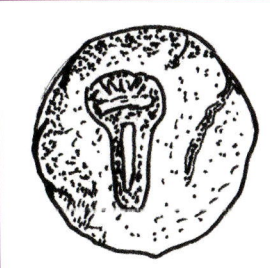

Fig.73. Lead trade weight of James I, with the letter "I" below a crown.

119

Lead Weights

Fig.74. Lead trade weight of William and Mary, "WR" below a crown. To the right the dagger of London and below the plumber's mark.

Pics.1a & b. Lead trade weight with unknown mark, possibly a bird in flight. An attempt has been made to produce a milled effect similar to a coin. The reverse shows an irregular shape.

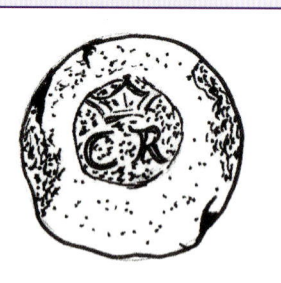

Fig.75. Lead trade weight of Charles I, "CR" below a crown within a circle.

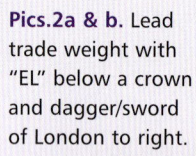

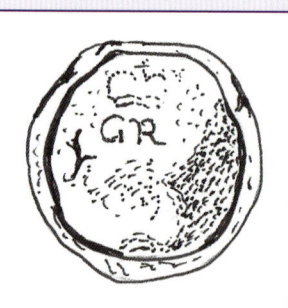

Pics.2a & b. Lead trade weight with "EL" below a crown and dagger/sword of London to right.

Fig.76. Lead trade weight, possibly George III with "GR" below a beaded crown, raised outer border.

Pics.3a & b. Medieval lead trade weight, unclear design, hole at top.

Pics.4a & b. Medieval lead trade weight, raised shield with roped edge.

Lead Weights

Pics.5a & b. Lead trade weight of either Henry VII or VIII, "h" below crown with possible plumber's mark above. The reverse has the initials of Elizabeth I "EL". This would indicate that the weight remained in use for a long time.

Pics.7a & b. Lead trade weight, post medieval, nine spoke wheel. The reverse shows a central hole.

Pics.6a & b. Lead trade weight with "WM" (William and Mary) below a crown, to the right the dagger/sword of London, and below the plumber's mark.

Pics.8a & b. Lead trade weight with circle of leaves. The reverse shows the letter "D".

121

Lead Weights

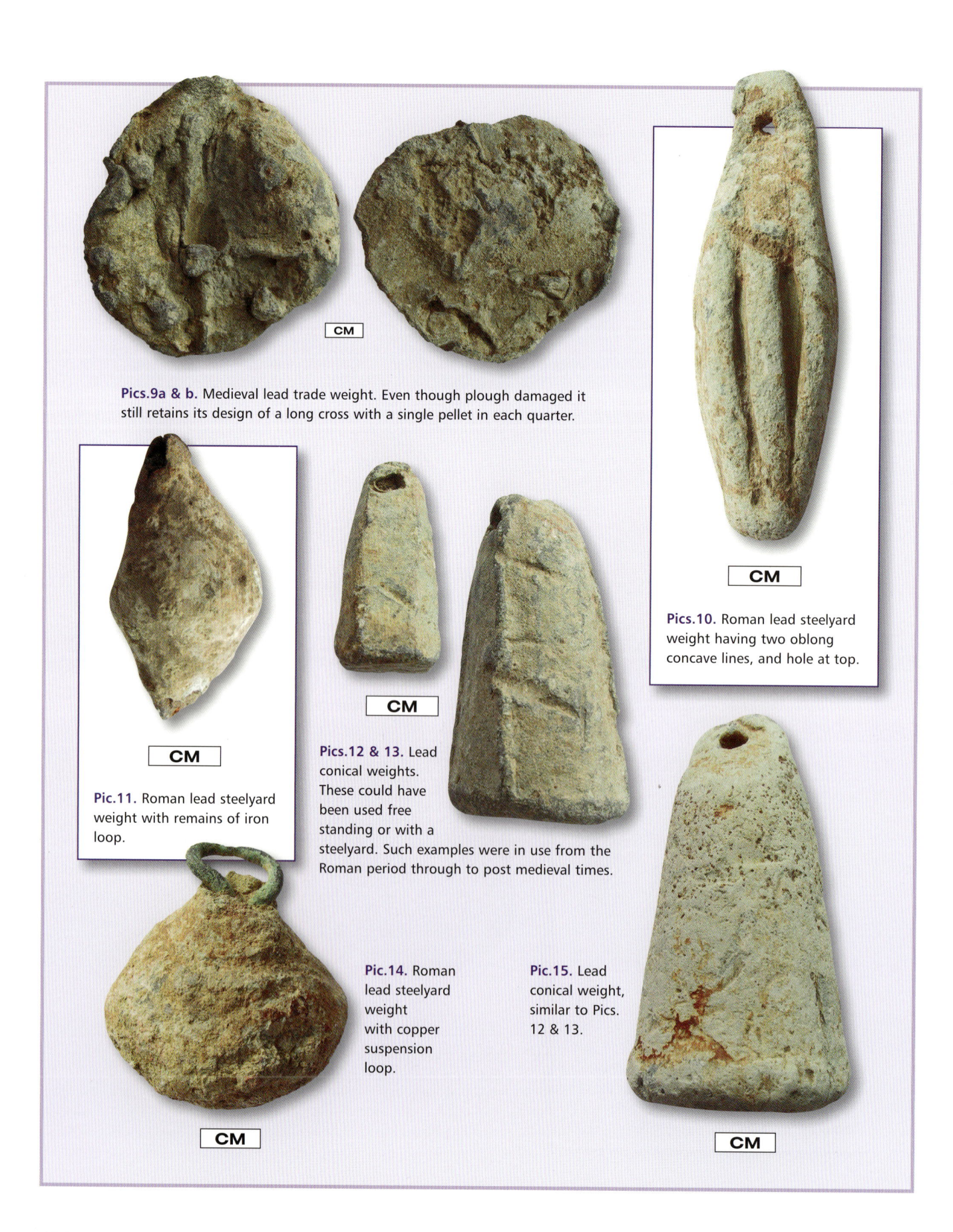

Pics.9a & b. Medieval lead trade weight. Even though plough damaged it still retains its design of a long cross with a single pellet in each quarter.

Pics.10. Roman lead steelyard weight having two oblong concave lines, and hole at top.

Pic.11. Roman lead steelyard weight with remains of iron loop.

Pics.12 & 13. Lead conical weights. These could have been used free standing or with a steelyard. Such examples were in use from the Roman period through to post medieval times.

Pic.14. Roman lead steelyard weight with copper suspension loop.

Pic.15. Lead conical weight, similar to Pics. 12 & 13.

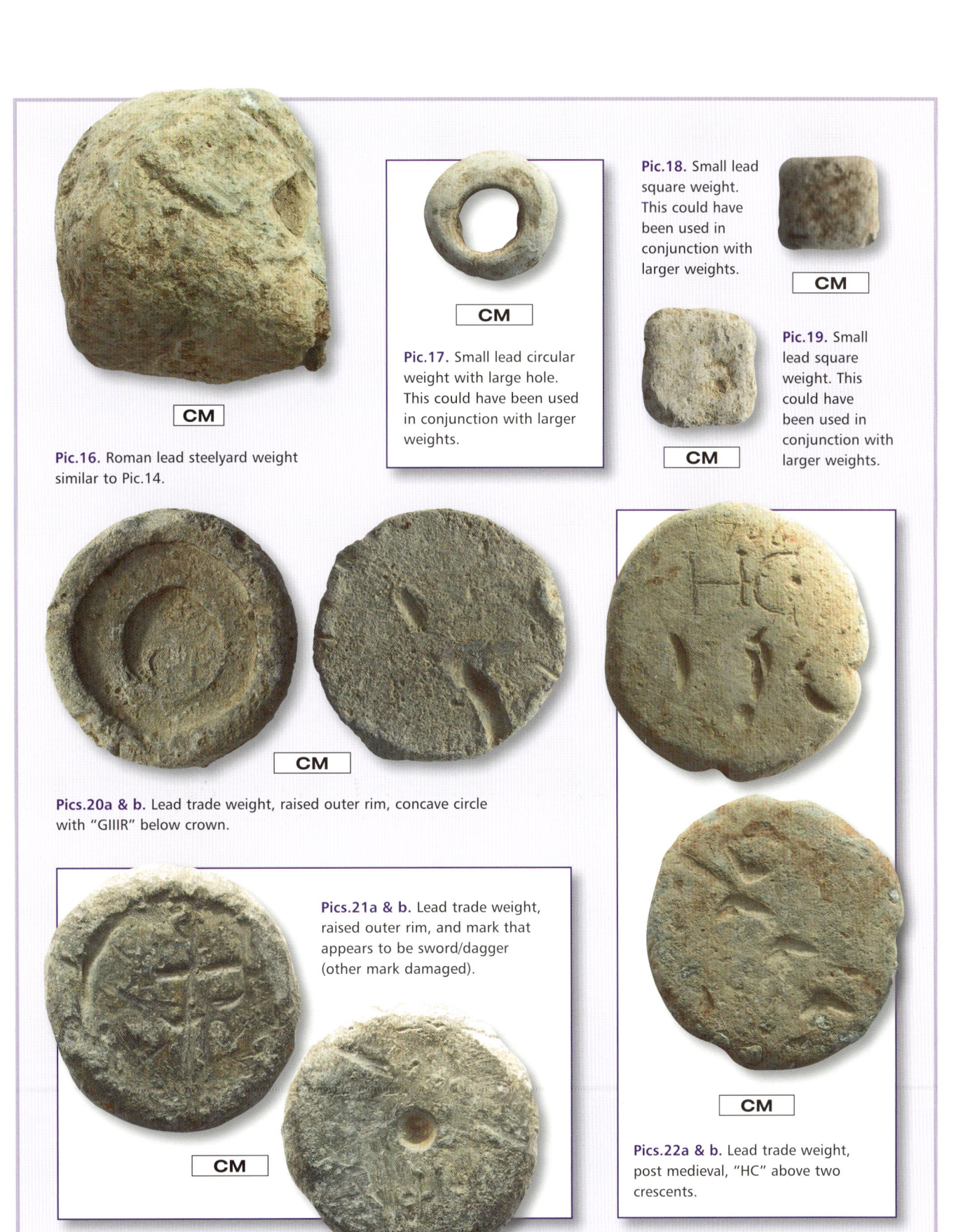

Lead Weights

Pic.16. Roman lead steelyard weight similar to Pic.14.

Pic.17. Small lead circular weight with large hole. This could have been used in conjunction with larger weights.

Pic.18. Small lead square weight. This could have been used in conjunction with larger weights.

Pic.19. Small lead square weight. This could have been used in conjunction with larger weights.

Pics.20a & b. Lead trade weight, raised outer rim, concave circle with "GIIIR" below crown.

Pics.21a & b. Lead trade weight, raised outer rim, and mark that appears to be sword/dagger (other mark damaged).

Pics.22a & b. Lead trade weight, post medieval, "HC" above two crescents.

123

Lead Weights

Pics.23a & b. Lead trade weight, raised rim, concave central hole.

Pics.24a & b. Lead trade weight, post medieval, three lined circles the outer having illegible lettering.

Pics.25a & b. Lead trade weight, pellets below a crescent.

Pics.26a & b. Lead trade weight showing two crosses.

Pics.27a & b. Medieval lead weight, long cross with radiate lines in each quarter.

Pic.28. Conical lead trade weight similar to Pics.12 & 13.

124

Lead Weights

Lead Trade Weights in Stan Raymond Collection (Not to Scale)

Pic.1. Medieval lead trade weight, three lions passant within a lined border.

Pic.2. Medieval lead trade weight, lion rampant (?) within a lined border.

Pic.3. Medieval lead trade weight, worn raised shield design surrounded by pellets.

Pic.4. Medieval lead trade weight, Cross of St Andrew within a lined border.

Pic.5. Medieval lead trade weight, distorted design within a lined border. The weight has a hole at the top and signs of rust where an iron suspension ring would once have been.

Pic.6. Medieval lead trade weight showing a crude fleur-de-lis.

Pic.7. Medieval lead trade weight, showing a well cast fleur-de-lis.

Pic.8. Medieval lead trade weight, the whole object being of fleur-de-lis shape.

125

Lead Weights

Various Marks Found on Lead Weights

1 & 2. Often described as a dagger or sword. The former supposedly representing the dagger used by the Mayor of London when stabbing Watt Tyler in 1381. However, it is also said that the sword was granted by Richard II as a reward for services to William Walworth. There is no documented evidence for either theory and the design, when used on weights, tends to vary in style.

3. "IR" below a crown representing King James II.

4. "I" below a crown representing King James I.

5. Shield of London (NB no dagger or sword).

6. Uncertain – possibly representing a bird in flight.

7. "EL" below a crown for Queen Elizabeth I.

8. Crown with no mark beneath, possibly 14th-15th century.

9. "C II R" below crown, King Charles II.

10. "AR" below crown, Queen Anne.

11. "A" below crown, Queen Anne.

12. "EL" below crown, Queen Elizabeth I.

13. "EL" below crown, Queen Elizabeth I.

14. "WM" below crown, King William and Queen Mary.

15. "CR" below crown, King Charles I.

16. Two shields within circle, one having the shield of London the other a harp, Commonwealth.

17. "C" below crown, King Charles I.

18. "h" below a crown, King Henry VII or VIII.

19. "WR" below a crown, King William III.

20. Plumbers Company mark.

21. Norwich (in Norfolk) mark.

22. "GR" below a dotted crown, King George III.

OTHER GREAT BOOKS IN THIS SERIES

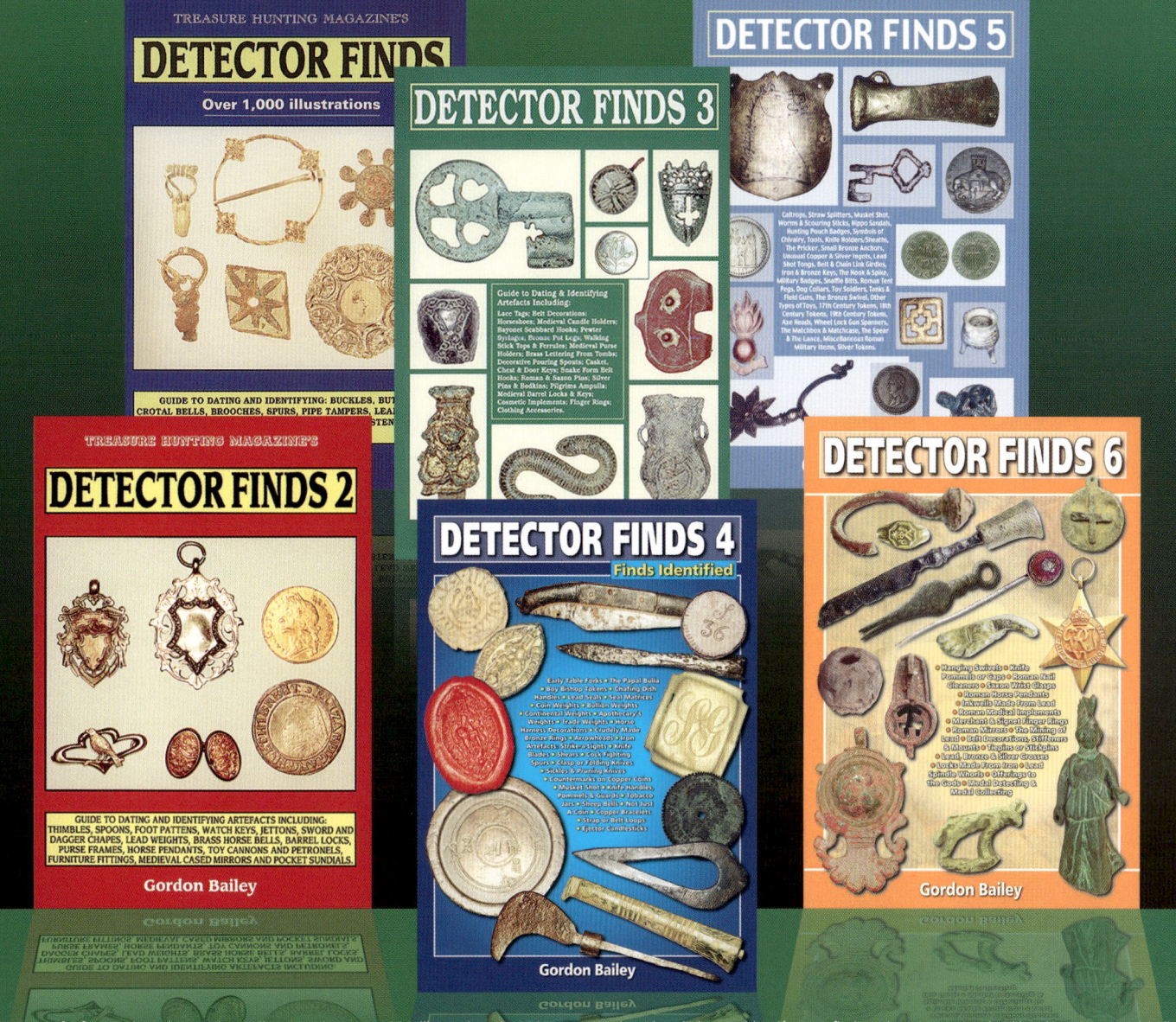

Detector Finds 1 written by Gordon Bailey contains over 1000 illustrations in 100 pages. Designed to help you identify your finds, volume 1 covers: buckles, buttons, crotal bells, brooches, spurs, pipe tampers, lead weights, cook clasps, lead tokens and hook fasteners. A4, 100 pages, £15.00 ISBN 1 897738 021

Detector Finds 2 again written by Gordon Bailey has 100 pages with hundreds of illustrations covering: thimbles, spoons, foot pattens, watch keys, jettons, sword & dagger chapes, lead weights, brass horse bells, barrel locks, purse frames, horse pendants, toy cannons & petronels, furniture fittings, medieval cased mirrors and pocket sundials. A4, 100 pages, £15.00 ISBN 1 897738 013

Detector Finds 3 by Gordon Bailey has 1000's of finds described and illustrated by means of clear line drawings and photographs. All new material. A large percentage of the photographs are now in full colour. "Exploded View" diagrams showing how and where artefacts such as clothing accessories were used. Gordon Bailey has been researching detector finds for over 20 years. This book gives you instant access to his vast wealth of experience. Over 30 individual categories of finds covered. Finds identified from all periods, Roman to Modern. A4, 96 pages, £15.00 ISBN 1 897738 226

Detector Finds 4 Another superb reference book from this popular author that will help you identify and value a wide range of finds. Finds Identified is extensively illustrated, in full colour, and contains all new material. Covering the period Medieval to Victorian (1100-1900) the contents include: Early Table Forks • The Papal Bulla • Boy Bishop Tokens • Chafing Dish Handles • Lead Seals • Seal Matrices • Coin Weights • Bullion Weights • Continental Weights • Apothecary's Weights • Trade Weights • Horse Harness Decorations • Crudely Made Bronze Rings • Arrow Heads • Strike-a-Lights • Knife Blades • Shears • Cock Fighting Spurs • Clasp or Folding Knives • Sickles & Pruning Knives • Countermarks on Copper Coins • Musket Shot • Knife Handles, Pommels & Guards • Tobacco Jars • Sheep Bells • Not Just A Coin • Copper Bracelets • Strap or Belt Loops • Ejector Candlesticks • Cannon Balls • Counterfeit Coins • Lead Tokens A4, 100 pages, £15.00 ISBN 1 897738 323

Detector Finds 5 contains hundreds of "new" colour illustrations of artefacts to help you identify, date and price your finds. Covering the stone age to Victorian times the chapter titles give you a clear indication of the wide appeal of DF5. As you can see there is significant coverage of tokens: Caltrops, Straw Splitters, Musket Shot, Worms & Scouring Sticks, Hippo Sandals, Hunting Pouch Badges, Symbols of Chivalry, Tools, Knife Holders/Sheaths, The Pricker, Small Bronze Anchors, Unusual Copper & Silver Ingots, Lead Shot Tongs, Belt & Chain Link Girdles, Iron & Bronze Keys, The Hook & Spike, Military Badges, Snaffle Bitts, Roman Tent Pegs, Dog Collars, Toy Soldiers, Tanks & Field Guns, The Bronze Swivel, Other Types of Toys, 17th Century Tokens, 18th Century Tokens, 19th Century Tokens, Axe Heads, Wheel Lock Gun Spanners, The Matchbox & Matchcase, The Spear & The Lance, Miscellaneous Roman Military Items, Silver Tokens. A4, 100 pages, £15.00 ISBN 1 897738 102

Detector Finds 6 Using over 660 clear illustrations of never seen before artefacts, to help you identify finds these are the topics included: • Hanging Swivels • Knife Pommels or Caps • Roman Nail Cleaners • Saxon Wrist Clasps • Roman Horse Pendants • Inkwells Made From Lead • Roman Medical Implements • Merchant & Signet Finger Rings • Roman Mirrors • The Mining of Lead • Belt Decorations, Stiffeners & Mounts • Tiepins or Stickpins • Lead, Bronze & Silver Crosses • Locks Made From Iron • Lead Spindle Whorls • Offerings to the Gods • Medal Detecting & Medal Collecting. A4, 112 pages, £15.00 ISBN 978 1 897738 313

01376 521900
www.greenlightpublishing.co.uk

GREENLIGHT publishing

GREENLIGHT

In **Reading Beaches** Ted Fletcher tells you how to be in the right place, at the right time and with the right detector and shows you how to identify the most productive search spots. This A5 title runs to 88 colour pages and contains over 50 illustrations.
A5, 88 pages, £8.00
ISBN 1 8977 38 153

Reading Land This title draws the readers attention to sites where people have congregated over the years, and where, naturally, losses of coins, jewellery etc have increased dramatically.
A5, 100 pages, £8.00 ISBN 1 8977 38 110

Reading Tidal Rivers This title shows you where to look on British & European tidal rivers for those elusive metallic artefacts that have been lost over the years.
A5, 84 pages, £8.00 ISBN 1 8977 38 080

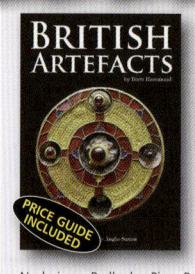

BUY ALL 3 SAVE £4.00

British Artefacts Volume 1 – Early Anglo-Saxon by Brett Hammond. The book contains 20 maps showing the distribution throughout Britain of various classes of objects and has 240 beautiful illustrations. **Contents:** Runes, Advice for Finders, Outline of the Early Anglo-Saxon Period, Art styles, Artefacts production & distribution, Ceramic production and Metal Artefacts. **Artefacts include:** Brooches, Buckles & Belt fittings, Clasps, Weapons & Fittings, Bowls & Vessels, Pendants, Belt Rings, Bracelets & Arm-rings, Chatelaines, Latch-lifters & Girdle-Hangers, Keys, Combs, Earrings, Finger-rings, Harness & Bridle Mounts, Neck-rings, Padlocks, Pins, Purse Mounts & Fire-steels, Pyxides, Spoons, Spurs, Tags, Metallic Threads, Toilet Sets, Tools and Weaving Equipment. The Non-Metallic Artefacts include Amber, Antler, Bone, Ceramics, Gemstones, Glassware, Horn, Ivory and Stone. A4, 132 pages, £15.00 ISBN 978 1 897738 351

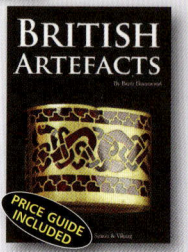

British Artefacts Volume 2 – Middle Saxon & Viking by Brett Hammond. This second volume covers the Middle Saxon material, including the impact th Vikings had on Anglo-Saxon life dur the period. **Contents:** The Middle Saxo Viking Period; Runes & Roman Script; A fact Production & Distribution; Art Sty Advice for Collectors; Advice for Finde Metal Artefacts – including brooch buckles, strap ends, fasteners & tags, p dants, mounts, rings & bracelets, weape & fittings, ecclesiastical & liturgical iter gaming pieces, weights, keys, tools & utensils; Non-Metallic Artefa Anglo Saxon & Viking Burials; Middle Saxon & Viking Kingdoms – lavishly illustrated with nearly 400 beautiful colour pictures and ma A4, 148 pages, £15.00 ISBN 1 897738 382

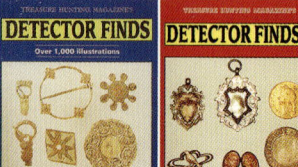

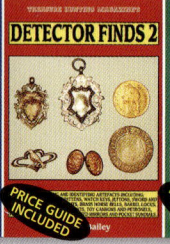

BUY ALL 6 AND SAVE £20.00

Detector Finds 1 written by Gordon Bailey contains over 1000 illustrations 100 pages. Designed to help you identify your finds, volume 1 covers: buckl buttons, crotal bells, brooches, spurs, pipe tampers, lead weights, cook clas lead tokens and hook fasteners.
A4, 100 pages, £15.00 ISBN 1 897738 021

Detector Finds 2 again written by Gordon Bailey has 100 pages w hundreds of illustrations covering: thimbles, spoons, foot pattens, watch ke jettons, sword & dagger chapes, lead weights, brass horse bells, barrel loc purse frames, horse pendants, toy cannons & petronels, furniture fittin medieval cased mirrors and pocket sundials.
A4, 100 pages, £15.00 ISBN 1 897738 013

Detector Finds 3 by Gordon Bailey has 1000's of finds described a illustrated by means of clear line drawings and photographs. All ne material. A large percentage of the photographs are now in full colo clothing accessories were used. Gordon Bailey has been researching detec finds for over 20 years. This book gives you instant access to his vast wealth of experience. Over 30 individual categories of finds covered. Finds identified from all periods, Roman to Modern. A4, 96 pages, £15.00 ISBN 1 897738 2

Detector Finds 4 Another superb reference book from this popular author that will help you identify and value a wide range of finds. Finds Identified is extensively illustrated, in full colour, all containing all new material. Covering the perio Medieval to Victorian (1100-1900) the contents include: Early Table Forks ● The Papal Bulla ● Boy Bishop Tokens ● Chafing Dish Handles ● Lead Seals ● Seal Matrices ● Coin Weights ● Bullion Weights ● Continental Weights ● Apothecary's Weight Trade Weights ● Horse Harness Decorations ● Crudely Made Bronze Rings ● Arrow Heads ● Strike-a-Lights ● Knife Blades ● Shears ● Cock Fighting Spurs ● Clasp or Folding Knives ● Sickles & Pruning Knives ● Countermarks on Copper Coins ● Musl Shot ● Knife Handles, Pommels & Guards ● Tobacco Jars ● Sheep Bells ● Not Just A Coin ● Copper Bracelets ● Strap or Belt Loops ● Ejector Candlesticks ● Cannon Balls ● Counterfeit Coins ● Lead Tokens A4, 100 pages, £15.00 ISBN 1 897738 3

Detector Finds 5 contains hundreds of "new" colour illustrations of artefacts to help you identify, date and price your finds. Covering the stone age to Victorian times the chapter titles give you a clear indication of the wide appeal of DF5. As y can see there is significant coverage of tokens: Caltrops, Straw Splitters, Musket Shot, Worms & Scouring Sticks, Hippo Sandals, Hunting Pouch Badges, Symbols of Chivalry, Tools, Knife Holders/Sheaths, The Pricker, Small Bronze Anchors, Unus Copper & Silver Ingots, Lead Shot Tongs, Belt & Chain Link Girdles, Iron & Bronze Keys, The Hook & Spike, Military Badges, Snaffle Bitts, Roman Tent Pegs, Dog Collars, Toy Soldiers, Tanks & Field Guns, The Bronze Swivel, Other Types of Toys, 1 Century Tokens, 18th Century Tokens, 19th Century Tokens, Axe Heads, Wheel Lock Gun Spanners, The Matchbox & Matchcase, The Spear & The Lance, Miscellaneous Roman Military Items, Silver Tokens. A4, 100 pages, £15.00 ISBN 1 897738 1

Detector Finds 6 the latest in the series from Gordon Bailey. Using over 660 clear illustrations of never seen before artefacts, to help you identify finds these are the topics included: ● Hanging Swivels ● Knife Pommels or Caps ● Roman N Cleaners ● Saxon Wrist Clasps ● Roman Horse Pendants ● Inkwells Made From Lead ● Roman Medical Implements ● Merchant & Signet Finger Rings ● Roman Mirrors ● The Mining of Lead ● Belt Decorations, Stiffeners & Mounts ● Tiepins or Stickp ● Lead, Bronze & Silver Crosses ● Locks Made From Iron ● Lead Spindle Whorls ● Offerings to the Gods ● Medal Detecting & Medal Collecting. A4, 112 pages, £15.00 ISBN 978 1 897738 313

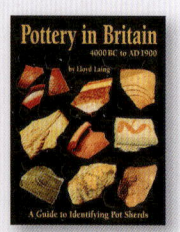

Pottery in Britain, a guide to identifying pot sherds, by Lloyd Laing, aims to provide an introductory guide to identifying some of the basic types of pottery that may be found and contains 178 illustrations, in the following sections: The potter's craft ● The study of pottery ● Prehistoric pottery – the Neolithic Period circa 4000-2000BC ● The Bronze Age circa 2000-700BC ● The Iron Age circa 700BC-AD43 ● The Roman Period AD43 – circ AD409 ● The Dark Ages & Early Medieval Period ● The Medieval Period – 11th-15th Centuries ● The 16th & 17th Centuries ● The 18th & 19th Centuries ● Glossary of terminology.
250mm x 190mm, 136 pages, £20.00 ISBN 1 897738 145

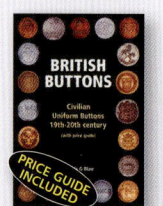

British Buttons by Dennis Blair An authoritative book compiled for collectors and those interested in the design of buttons. This book is an ideal reference work for identification of button finds. It contains 375 button examples reproduced in colour, including Livery and Royal Court buttons as well as General issues; there is also a chapter upon Button Making. **The chapter headings are:** 1 – General Overview 2 – Differentiations & Updating 3 – Livery Buttons 4 – Royal Court Buttons 5 – Collecting Themes 6 – Button Making & Bookmarks
A5, 92 pages, £8.00 ISBN 1 8977 38 04 8

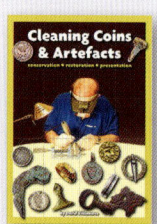

Cleaning Coins & Artefacts (conservation ● restoration ● pre entation) by David Villanueva, sets o to show you what you can safely do clean & preserve metal detector finds. T chapter headings give you an idea of t coverage of this title: Introduction ● The Field ● Map Reading ● Safe Stora ● Identification and Assessment ● Int duction To Cleaning Finds ● Mechanic Cleaning ● Electrolysis ● Chemical Clea ing and Conservation ● Repair, Restorati and Replication ● Photographing Your Finds ● Storage And Disp ● The Treasure Act ● Bibliography and Suppliers.
A5, 116 pages, £12.00 ISBN 978 1 897738 337

Buttons & Fasteners 500 BC-AD 1840 by Gordon Bailey. With over 1,000 high quality colour photographs, this book allows the identification and dating of metal buttons from the Iron-age to early 19th century. It also covers Iron Age toggles, Saxon-Tudor hook fasteners, and ring brooches. **CONTENTS:** Excavated Metal Buttons ● Bronze Age ● Iron Age ● Roman ● Saxon & Viking ● Medieval (11th & 12th Centuries) ● Medieval (13th-15th Centuries) ● Late 15th-17th Centuries ● 17th Century ● 18th-Early 19th Century ● Celtic Toggles ● Saxon, Viking & Tudor Hook Fasteners ● Ring Brooches 250mm x 190mm, 100+ pages, £16.00 ISBN 1 897738 218

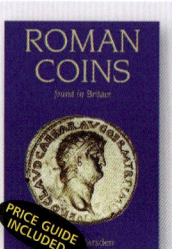

Roman Coins found in Britain Roman coins were used in Britain for nearly 400 years and are common finds in the soil of this country. **Contents:** Introduction to Roman coins and their identification ● Coin legends and understanding the inscriptions on Roman coins ● Portraits and propaganda ● Mints: Differences between mints, mintmarks in the later empire ● Roman coins in the earlier empire, up to 238 ● The radiate coinages, 238-296 ● The fourth century ● The end of Roman coinage in Britain ● Contemporary forgeries in Roman Britain ● Treatment of coins, preservation, cataloguing etc.
A4, 108 pages, £16.00 ISBN 1 8977 38 06 4

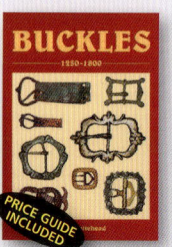

Buckles 1250-1800, written by Ross Whitehead, contains over 800 illustrated buckles (mostly in colour) with full descriptions and background text. A unique classification format using shape rather than type or period, aids identification. Contents include; buckle manufacture, single looped buckles, buckles with integral plates, clasp fastners, annular buckles, rectangular and trapezoidal buckles, asymmetrical buckles, two piece buckles and finally buckles as jewellery. A4, 128 pages, £16.00 ISBN 1 8977 38 17x

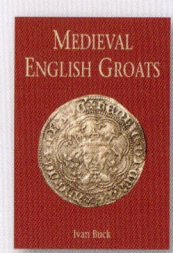

Medieval Englis Groats. This is th definitive referen work on Englis Groats. Written Ivan Buck, it cove the groat from introduction in the reign of Edward (1272-1307) rig up to the end of th Tudors in the ear 17th century. Essenti reading – this work helps to identify the vario types of groat and the major varieties. There a over 400 colour illustrations in the text and number of scarce and rare coins are illustrate for the first time. In many cases the informatio provided can be applied to the parallel series half groats.
A4, 68 pages, £16.00 ISBN 1 8977 38 420

---

order online www.greenlightpublishing.co.uk ☎ **orders 01376 521900**

ORDER FORM – Send to: Greenlight Publishing, 119 Newland Street, Witham, Essex CM8 1WF ENGLAND

Please supply:

British Artefacts Volume 1 .... @ £15.00 UK post free ☐	Detector Finds 5 .... @ £15.00 UK post free ☐	
British Artefacts Volume 2 .... @ £15.00 UK post free ☐	Detector Finds 6 .... @ £15.00 UK post free ☐	
Buttons & Fasteners 500BC-AD1840... @ £16.00 UK post free ☐	All 1-6 Detector Finds Books .... @ £70.00 UK post free ☐	
British Buttons .... @ £8.00 UK post free ☐	Medieval English Groats .... @ £16.00 UK post free ☐	
Buckles 1250-1800 .... @ £16.00 UK post free ☐	Pottery in Britain .... @ £20.00 UK post free ☐	
Cleaning Coins & Artefacts .... @ £12.00 UK post free ☐	Reading Beaches .... @ £8.00 UK post free ☐	
Detector Finds 1 .... @ £15.00 UK post free ☐	Reading Land .... @ £8.00 UK post free ☐	
Detector Finds 2 .... @ £15.00 UK post free ☐	Reading Tidal Rivers .... @ £8.00 UK post free ☐	
Detector Finds 3 .... @ £15.00 UK post free ☐	All 3 'Reading' books .... @ £20.00 UK post free ☐	
Detector Finds 4 .... @ £15.00 UK post free ☐	Roman Coins Found in Britain .... @ £16.00 UK post free ☐	

ALLOW 3-5 days for delivery

Payment enclosed: £ _____

cheques should be made payable to Greenlight Publishing
OVERSEAS PAYMENTS BY CREDIT CARD ONLY

Card security code
(last 3 digits on signature strip)

Send to: NAME ....
ADDRESS ....
....POSTCODE ....
TEL .... EMAIL ....

Expiry Date ....
Maestro card issue No ....
Valid from ....
Sign ....

GREENLIGHT

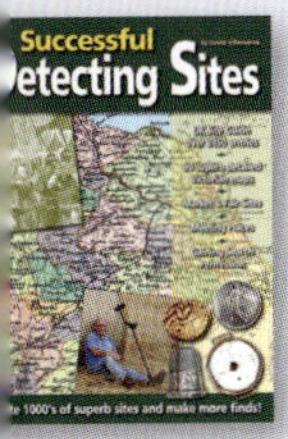

Successful Detecting Sites by David Villanueva contains over 2450 UK site entries. Using rare 18th & 19th century sources, David Villanueva has drawn on over 30 years experience in metal detecting and historical research to compile this exciting guide to thousands of potentially successful detecting sites throughout the United Kingdom, with histories stretching back ndreds or even thousands of years. He explains clearly how to ierate a host of successful detecting sites from every place in guide, which will keep your finds bag overflowing for years come. And to lead you to these Sites, there is a wealth of uable information included together with superb facsimiles 92 highly detailed Victorian maps covering every UK county you get a complete antique county atlas as well. **Contents:** tory of Markets and Fairs in Britain • The Siting of Markets and rs • Finds from Market and Fair Sites • Finds from the Routes • en-Air Political Meetings • The Siting of Meeting Places • Finds m a Hundred Court Site • Practical Map Reading • Finding Sites • County Atlas and Site Guide for England and Wales • unty Atlas and Site Guide for Scotland • County Atlas and Site ide for Northern Ireland & Offshore Islands • Gaining Search mission – The Project Approach • Bibliography and Sources • ide of Practice
0mm x 190mm, 238 pages, £20.00 ISBN 978 1 897738 306

BUY BOTH AND SAVE £5.00

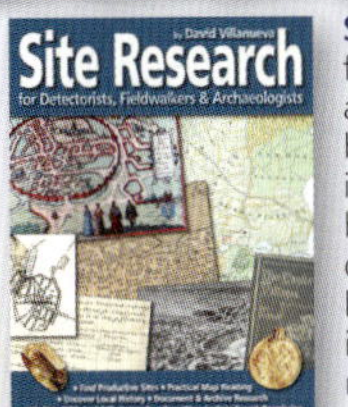

Site Research Why should one field be productive of finds year after year and yet the next field be totally barren? The answer is past human activity, and this book shows through map and document research, how to locate such activity. Profusely illustrated with examples of maps and documents, and finds resulting from the suggested research methods. Although written mainly for detectorists, this book will also be of interest and help to fieldwalkers, local historians and archaeologists. David Villanueva has over 30 years of experience in metal detecting and research and has been responsible for seven reported finds of Treasure. Using this book you will start to acquire more productive sites and as a result start to make better finds. ***Chapter titles:*** • Using Archives, Libraries & Computers • County Maps • Ordnance Survey Maps • Practical Map Reading • Town Plans • Road Maps • Road, River, Canal & Railway Construction Maps • Enclosure & Tithe Maps • Estate Maps • Sea Charts • Aerial Photographs, Maps & Surveys • Local Histories • Guide to County Histories • Domesday Book • Gaining Search Permission • Search Agreements • Living with the Treasure Act • Code of Practice • Bibliography & Sources
250mm x 190mm, 160 pages, £20.00
ISBN 1 897738 285

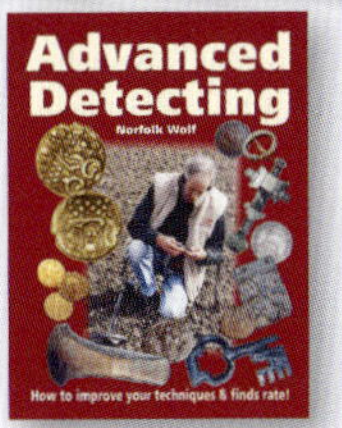

Advanced Detecting by John Lynn, better known as the "Norfolk Wolf", one of the country's leading authorities on metal detecting. This book is a must have for the detectorist wishing to improve his or her skills. It explains in an easy-to-read style every problem or situation that a detectorist is likely to encounter and, more importantly, what causes them and how they can be overcome. **Chapter titles are:** Understanding your Detector • Mineralisation & Ground Effect • Meters or Audio • Discrimination and the Conductivity of Metals • The Functions of Sensitivity & All Metal • Mindset, Experience, Confidence & Concentration • Starting From Scratch on a New Field • Time-Out But Not to Smell The Flowers • Sweep, Stem Lengths & Pace Lengths • Signals • The Best & Worst Times to Detect & Different Surfaces • Identifying Pottery • Recovery & Response Speed • Bits & Pieces
250mm x 190mm, 108 pages, £16.00 ISBN 1 897738 250

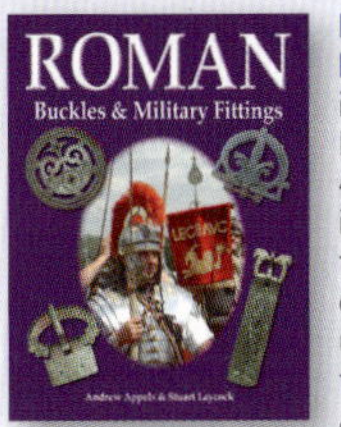

Roman Buckles & Military Fittings The history of Britain is intimately tied up with the Roman army which for almost 400 years kept most of this island Roman. Over the centuries that the Roman army occupied Britain, its' soldiers used a bewildering variety of fittings. In this book, Laycock and Appels set out to document and identify many of the items of Roman military kit encountered today by detectorists and archaeologists and set them in their historical and military context. This text is lavishly illustrated with over 900 full colour photographs of surviving Roman military kit, most of them never before published. Some of these items are unique. Many of them are rare. This is a resource for detectorists, archaeologists, museum staff, collectors and re-enactors alike, and will be of interest also to **anyone with a more general interest in the Roman military. *Chapter titles:*** Early Empire Buckles • Dolphin Buckles • Dragon Buckles • Bird Buckles • Horse head buckles • Lion Buckles • Strap ends • Belt stiffeners • Belt Plates • Helmet Fittings • Sword and Dagger Fittings • Armour Fittings • Shield Fittings • Apron Fittings • Horse harness Fittings • Roman military glossary • Bibliography
250mm x 190mm, 284 pages, £20.00
ISBN 1 897738 290

BUY ALL 3 AND SAVE £8.00

Celtic & Roman Artefacts by Nigel Mills has over 450 beautiful colour illustrations. **Contents:** • Bronze & Iron Age Artefacts • Fibula Brooches • Plate, Crossbow & Early Saxon Brooches • Buckles & Military Equipment • Locks, Keys & Knife Handles • Spoons, Cosmetic Grinders, Medical Implements & Seal Boxes • Jewellery • Cube Matrices, Lead Seals and Gaming Pieces • Pottery & Bronze ensils • Steelyard Weights & Bronze Mounts • Figurines & otive Objects • Chart of Roman Gods • Select Bibliography Full price guide for every item in two grades of condition.
4, 152 pages, £16.00 ISBN 1 8977 38 37 4

Medieval Artefacts An indispensible reference work, 116 pages, all colour, price guide, with over 300 beautiful illustrations spanning the period 1066-1500. **Contents include:** Introduction (Mudlarking and Historical Background), Buckles, Strap-ends, Seal Matrices, Thimbles, Pilgrim Badges, Finger Rings, Brooch Buckles, Buttons & Pins, Heraldic Pendants, Keys, Locks & Weights, Spoons, Knives Pottery, Gaming, Purses & Papal Bullas, Sporting & Hunting (inc Spurs, Arrowheads, Daggers, Sword Pommels & Chapes), Figurines & Church Vessels
A4, 116 pages, £16.00 ISBN 1 8977 38 27 7

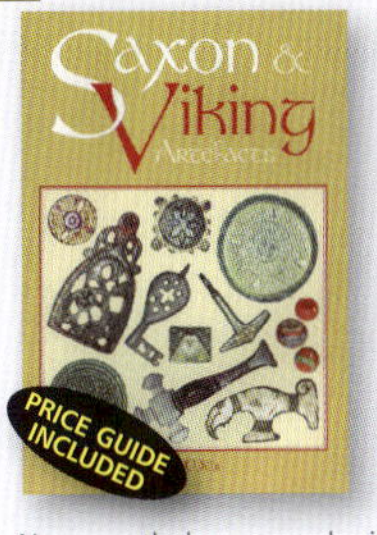

Saxon & Viking Artefacts covers the period from the 6th to 11th centuries and – together with "Celtic & Roman Artefacts" and "Medieval Artefacts" (also by Nigel Mills) – completes the historical series covering artefacts from the Bronze Age to Tudor times. Illustrated in full colour and with over 250 superb photographs of individual objects, it encompasses the full spectrum of everyday items in use in Anglo-Saxon England in chronological sequence. The selection of illustrations has been built up over a period of 15 years from various collections. The objects covered include: buckles, strap ends, pins, cruciform brooches, disc brooches, animal brooches, jewellery, beads, stirrup mounts, wrist clasps, dress hooks, keys, knives, tweezers, weights, gaming counters, and weapons. There is also a Norse mythology genealogical chart of the gods. There are additional notes and advice for collectors. The book is an invaluable reference work for collectors, dealers, museums, and archaeologists. Full price guide for every item in two grades of condition.
A4, 108 pages, £16.00 ISBN 1 8977 38 05 6

BUY ALL 3 AND SAVE £8.00

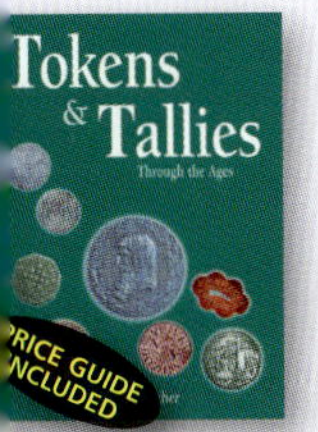

Tokens & Tallies through the ages by Edward Fletcher. Over 400 token illustrations from ancient to early 20th century. **CHAPTER HEADINGS:** Ancient Tallies & Tokens • Early English Tokens & Imitations • Medieval Jettons • Medieval Tokens • 17th Century Tokens • 18th Century Tokens • Communion Tokens • 19th Century base Metal & Silver Tokens • Unofficial Farthings & Other 19th Century Advertising Issues • Pub Checks • Work, Play & Games • Market Traders' Tallies • Tokens & Tallies In 19th & Early 20th Century Agriculture • Control In The Workplace
50mm x 190mm, 100 pages, £16.00 ISBN 1 897738 160

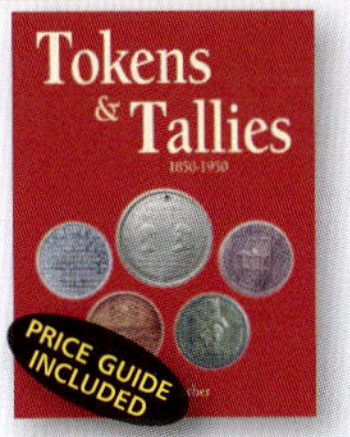

Tokens & Tallies 1850-1950 by Edward Fletcher. Looks in more detail at this period, with over 600 illustrations. **Chapter headings:** Introduction • Makers • Advertising • Regal Images • Calendars • Agricultural Tallies • Co-op • Bonus • Club • Pub • Temperance • Refreshment • Canteen • Teaching • Market • Gaming • Newspapers • Transport • Industry • Countermarks • Military & Wartime • Plastics • Further Reading
250mm x 190mm, 100+ pages, £16.00
ISBN 1 897738 196

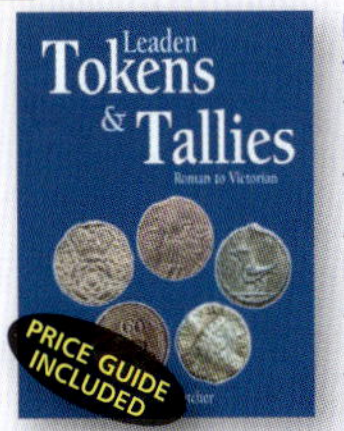

Leaden Tokens & Tallies – Roman to Victorian by Edward Fletcher. Over 780 illustrations of lead tokens. **CHAPTER HEADINGS:** Introduction • Acknowledgements • A Note To Paranumismatists • In The Beginning • The Anglo-French Connection • Early English Tokens • The Boy Bishop Phenomenon • Tokens And Tallies After 1400 • Tudor Tokens and Tallies • Post-Tudor Developments • Agricultural Tallies: 17th-19th century • Communion Tokens • Shycocks • Moulds • A Brief Look At Cloth & Bag Seals • Collecting & Researching • Some Puzzling Pieces
250mm x 190mm, £16.00 ISBN 1 897738 269

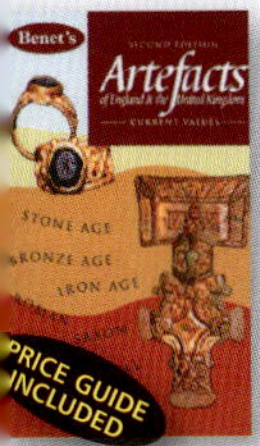

Benet's Artefacts second edition – contains over 600 pages. This superb artefact identifyer and price guide ranges from the Stone Age through to the Tudor period. With over 2000 beautifully illustrated artefacts, this hardback book is produced in full colour throughout. Benet's is a visual guide for identification purposes and to market prices. The prices quoted are based upon first hand knowledge nd experience of the antiquities market as well as knowledge f the actual prices obtained on many of the items shown. his knowledge has been combined with consultation with ther experienced and respected dealers in antiquities.
20mm x 140mm hardback, £20.00 616 pages
SBN 0 953617 211

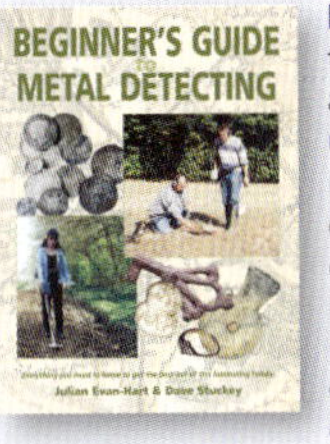

Beginner's Guide to Metal Detecting written by Julian Evan-Hart and Dave Stuckey is aimed at both those who are thinking of taking up this fascinating hobby and those who have recently started detecting but as yet have little experience. The title is well illustrated, in full colour, and the chapter headings give you a clear indication of the content. • The Different Types Of Detectors Available • Detecting Accessories • Where To Detect • Researching Potential Sites • Gaining Search Permission • Fieldwalking & What To Look For • Search Techniques & Methods • Examples Of Detector Found Artefacts • A Positive Attitude From Archaeology • Identifying, Recording, Cleaning, Storing & Displaying Finds • Making A Connection With Your Finds & Creating Time Lines • Further Reading & Assistance • Further Information • Glossary Of Terms
250mm x 190mm, 92 pages, £12.00 ISBN 1 897738 188

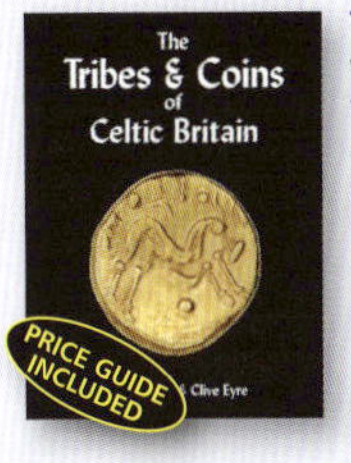

The Tribes & Coins of Celtic Britain The Celts left no written records and the only historical accounts we have of them derive mainly from Roman writers. As little as 30 years ago many mysteries – and misconceptions – still existed as to the Celtic tribes of Britain and their kings. But thanks to metal detecting finds and the Celtic Coin Index, far more is now known. In this book Rainer Pudill draws on his own experience as a collector – and this new knowledge – to present the latest thinking and facts on the Celts and their coins. **The contents include:** The Celtic Pantheon • Mercenaries & The First Celtic Coins • Iron Bars & Ring Money • Caesar's Expeditions To Britain • The Celtic Tribes Of Britain & Early Celtic Coinage • Pedigree • The British Policy Of Augustus & His Successors • The Coinages Of Cunobelin • Epaticcus • The Coinage Of Verica • The Invasion Of Britain • Resistance & Rebellion Against The Roman Occupation • The Final Celtic Coins Of Britain • The "Conquest Of The Rest" • Time Line. Contains over 300 illustrations and a price guide.
250mm x 190mm £16.00 ISBN 1 897738 242

order online www.greenlightpublishing.co.uk MasterCard VISA DELTA Maestro ☎ **orders 01376 521900**

ORDER FORM – Send to: Greenlight Publishing, 119 Newland Street, Witham, Essex CM8 1WF ENGLAND

ALLOW 3-5 days for delivery

Please supply:

Advanced Detecting	@ £16.00	UK post free	☐
Beginner's Guide to Metal Detecting	@ £12.00	UK post free	☐
Benet's Artefacts 2	@ £20.00	UK post free	☐
Celtic & Roman Artefacts	@ £16.00	UK post free	☐
Medieval Artefacts	@ £16.00	UK post free	☐
Saxon & Viking Artefacts	@ £16.00	UK post free	☐
All 3 Artefact books	@ £40.00	UK post free	☐
Roman Buckles & Military Fittings	@ £20.00	UK post free	☐
Site Research	@ £20.00	UK post free	☐
Successful Detecting Sites	@ £20.00	UK post free	☐
Successful Detecting Sites & Site Research	@ £35.00	UK post free	☐
The Tribes & Coins of Celtic Britain	@ £16.00	UK post free	☐
Leaden Tokens & Tallies	@ £16.00	UK post free	☐
Tokens & Tallies 1850-1950	@ £16.00	UK post free	☐
Tokens & Tallies through the ages	@ £16.00	UK post free	☐
All 3 Tokens & Tallies books	@ £40.00	UK post free	☐

Payment enclosed: £ ______

cheques should be made payable to Greenlight Publishing
OVERSEAS PAYMENTS BY CREDIT CARD ONLY

MasterCard ☐ VISA ☐ DELTA ☐ Maestro ☐

Card security code (last 3 digits on signature strip)

Expiry Date ....

Maestro card issue No ....

Valid from ....

Sign ....

Send to: NAME ....

ADDRESS ....

.... POSTCODE ....

TEL .... EMAIL ....

Treasure hunting

BRITAIN'S BEST SELLING METAL DETECTING MAGAZINE

Love metal Detecting? Love Treasure Hunting Magazine!
Published monthly and packed full of information about the fantastic hobby of metal detecting

Available in WH Smiths, all good detector retailers or direct from the publisher. Visit our website to see special subscriber offers or to request a free sample copy.

GREENLIGHT publishing

www.treasurehunting.co.uk or call 01376 521900